*f*P

ALSO BY ILYA PRIGOGINE

Order Out of Chaos

From Being to Becoming

Exploring Complexity

The End of
CERTAINTY

Time, Chaos, and the
New Laws of Nature

ILYA PRIGOGINE

In collaboration with Isabelle Stengers

THE FREE PRESS

New York   London   Toronto   Sydney   Singapore

THE FREE PRESS
A Division of Simon & Schuster Inc.
1230 Avenue of the Americas
New York, NY 10020

First Free Press Edition 1997
Published by arrangement with Éditions Odile Jacob, Paris, France

THE FREE PRESS and colophon are trademarks
of Simon & Schuster Inc.

Designed by Carla Bolte

Manufactured in the United States of America

10   9   8   7   6   5   4   3   2   1

Library of Congress Cataloging-in-Publication Data
Prigogine, I. (Ilya)
    [La fin des certitudes.   English]
    The end of certainty: time, chaos, and the new laws of nature/Ilya
Prigogine; in collaboration with Isabelle Stengers.
      p.   cm.
    Includes bibliographical references and index.
    1. Science—Philosophy.   2. Space and time.   3. Chaotic behavior
in systems.   4. Natural history.   I. Stengers, Isabelle.
II. Title.
Q175.P7513   1997
530.11—dc21   97–3001
                                                  CIP
ISBN   0–684–83705–6

Original French edition entitled *La Fin des Certitudes,* by Ilya Prigogine in
collaboration with Isabelle Stengers, published by Éditions Odile Jacob

CONTENTS

ACKNOWLEDGMENTS

This book has had a somewhat unusual history. Originally, Isabelle Stengers and I had intended to translate our book *Entre le Temps et l'Eternité* (Between Time and Eternity) into English.[1] We had already prepared several versions, one of which appeared in German, and another in Russian.[2] But at the same time, we were making important progress in the mathematical formulation of our approach. As a result, we abandoned our translation of the original book and decided to write a new version, which has recently appeared in French.[3] Isabelle Stengers has asked not to be designated as a coauthor of this new presentation, but only as my collaborator. Although I felt obliged to respect her wishes, I would like to stress that without her, this book would never have been written. I am most grateful for her assistance.

This text is the result of decades of work by both the Brussels and Austin groups. While the physical ideas had been clear for a considerable length of time, their precise mathematical formulation has emerged only during the past few years.[4] I express my gratitude to the young, enthusiastic coworkers who have been instrumental in defining the new approach to the nature of time that forms the basis of this book, especially Ioannis Antoniou (Brussels),

Dean Driebe (Austin), Hiroshi Hasegawa (Austin), Tomio Petrosky (Austin), and Shuichi Tasaki (Kyoto). I would also like to mention my old group in Brussels, who laid the foundations that made further progress possible. My thanks to Radu Balescu, Michel de Haan, Françoise Henin, Claude George, Alkis Grecos, and Fernand Mayné. Unfortunately, Pierre Résibois and Léon Rosenfeld are no longer with us.

The work presented in this book could not have been accomplished without the support of a number of organizations. I especially want to thank the Belgian Communauté Française, the Belgian federal government, the International Solvay Institutes (Brussels), the United States Department of Energy, the European Union, and the Welch Foundation (Texas).

English is not my native language, and I am much obliged to Dr. E. C. George Sudarshan and Dr. Dean Driebe, both from the University of Texas at Austin, and David Lortimer (London), who have read the text with great care. I also thank my French publisher, Odile Jacob, who encouraged me to write this new book, and Stephen Morrow, my editor in the United States, as well as Judyth Schaubhut Smith, for their help in preparing the English edition.

I believe that we are at an important turning point in the history of science. We have come to the end of the road paved by Galileo and Newton, which presented us with an image of a time-reversible, deterministic universe. We now see the erosion of determinism and the emergence of a new formulation of the laws of physics.

Ilya Prigogine

AUTHOR'S NOTE

I have tried to make this book a readable, self-contained account accessible to general readers. However, especially in chapters 5 and 6, I decided to go into more technical detail because the findings I have presented deviate significantly from traditional views. In spite of the fact that this volume is the result of decades of work, many questions still await answers. But taking into consideration the finite life span of each of us, the fruits of my labors are shown such as they are today. I invite readers not on a visit to an archaeological museum, but rather on an adventure in science in the making.

Introduction

A NEW RATIONALITY?

E arlier this century in *The Open Universe: An Argu-ment for Indeterminism,* Karl Popper wrote, "Common sense inclines, on the one hand, to assert that *every* event is caused by some preceding events, so that every event can be explained or predicted . . . . On the other hand, . . . common sense attributes to mature and sane human persons . . . the ability to choose freely between alternative possibilities of acting."[1] This "dilemma of determinism," as William James called it, is closely related to the meaning of time.[2] Is the future given, or is it under perpetual construction? A profound dilemma for all of mankind, as time is the fundamental dimension of our existence. It was the incorporation of time into the conceptual scheme of Galilean physics that marked the origins of modern science.

This triumph of human thought is also at the root of the main problem addressed by this book: the denial of what has been called the *arrow of time.* As is well known, Albert Einstein often asserted, "Time is an illusion." In-

1

deed time, as described by the basic laws of physics, from classical Newtonian dynamics to relativity and quantum physics, does not include any distinction between past and future. Even today, for many physicists it is a matter of faith that as far as the fundamental description of nature is concerned, there is no arrow of time.

Yet everywhere—in chemistry, geology, cosmology, biology, and the human sciences—past and future play different roles. How can the arrow of time emerge from what physics describes as a time-symmetrical world? This is the *time paradox,* one of the central concerns of this book.

The time paradox was identified only in the second half of the nineteenth century after the Viennese physicist Ludwig Boltzmann tried to emulate what Charles Darwin had done in biology in an effort to formulate an evolutionary approach to physics. The laws of Newtonian physics had long since been accepted as expressing the ideal of objective knowledge. As they implied the equivalence between past and future, any attempt to confer a fundamental meaning on the arrow of time was resisted as a threat to this ideal. Isaac Newton's laws were considered final in their domain of application, somewhat the way quantum mechanics is now considered to be final by many physicists. How then can we introduce unidirectional time without destroying these amazing achievements of the human mind?

Since Boltzmann, the arrow of time has been relegated to the realm of phenomenology. We, as imperfect human observers, are responsible for the difference between past and future through the approximations we introduce in our description of nature. This is still the prevailing scientific wisdom. Certain experts lament that we stand before an unsolvable mystery for which science can provide no answer. We believe that this is no longer the case because of two recent developments: the spectacular growth of nonequilibrium physics and the dynamics of unstable systems, beginning with the idea of chaos.

Over the past several decades, a new science has been born, the *physics of nonequilibrium processes*, and has led to concepts such as *self-organization* and *dissipative structures*, which are widely used today in a large spectrum of disciplines, including cosmology, chemistry, and biology, as well as ecology and the social sciences. The physics of nonequilibrium processes describes the effects of unidirectional time and gives fresh meaning to the term irreversibility. In the past, the arrow of time appeared in physics only through simple processes such as diffusion or viscosity, which could be understood without any extension of the usual time-reversible dynamics. This is no longer the case. We now know that irreversibility leads to a host of novel phenomena, such as vortex formation, chemical oscillations, and laser light, all illustrating the essential *constructive* role of the arrow of time. Irreversibility can no longer be identified with a mere appearance that would disappear if we had perfect knowledge. Instead, it leads to coherence, to effects that encompass billions and billions of particles. Figuratively speaking, matter at equilibrium, with no arrow of time, is "blind," but with the arrow of time, it begins to "see." Without this new coherence due to irreversible, nonequilibrium processes, life on earth would be impossible to envision. The claim that the arrow of time is "only phenomenological," or subjective, is therefore absurd. We are actually the children of the arrow of time, of evolution, not its progenitors.

The second crucial development in revising the concept of time was the formulation of the physics of unstable systems. Classical science emphasized order and stability; now, in contrast, we see fluctuations, instability, multiple choices, and limited predictability at all levels of observation. Ideas such as chaos have become quite popular, influencing our thinking in practically all fields of science, from cosmology to economics. As we shall demonstrate, we can now extend classical and quantum physics to include instability and chaos. We are then able to obtain a formulation of the laws of nature appropriate for the description of our evolving universe, a description that contains the arrow of time, since past and future no longer play symmetrical roles. In the classical view—and here we include quantum mechanics and relativity—laws of nature express certitudes. When appropriate initial conditions are given, we can predict with certainty the future, or "retrodict" the past. Once instability is included, this is no longer the case, and the meaning of the laws of nature changes radically, for they now express possibilities or probabilities. Here we go against one of the basic traditions of Western thought, the belief in certainty. As stated by Gerd Gigerenzer et al. in *The Empire of Chance*, "Despite the upheavals in science in the over two millennia separating Aristotle from the Paris of Claude Bernard, they shared at least one attitude of faith: Science was about causes, not chance. Kant even promoted universal causal determinism to the status of a necessary condition of all scientific knowledge."[3]

There were, however, dissenting voices. The great physicist James Clerk Maxwell spoke of a "new kind of knowledge" that would overcome the prejudice of determinism.[4] But, on the whole, the prevailing opinion was that probabilities were states of mind rather than states of the world. This is so even today in spite of the fact that quantum mechanics has included statistical concepts in the core of physics. But the fundamental object of quantum mechanics, the *wave function*, satisfies a deterministic, time-reversible equation. To introduce probability and irreversibility, the orthodox formulation of quantum mechanics requires an observer.

Through his measurements, the observer would bring irreversibility to a time-symmetric universe. Again, as in the time paradox, we would be responsible in some sense for the evolutionary patterns of the universe. This role of the observer, which gave quantum mechanics its subjective flavor, was the main reason that prevented Einstein from endorsing quantum mechanics, and it has since led to unending controversies.

The role of the observer was a necessary concept in the introduction of irreversibility, or the flow of time, into quantum theory. But once it is shown that instability breaks time symmetry, the observer is no longer essential. In solving the time paradox, we also solve the quantum paradox, and obtain a new, realistic formulation of quantum theory. This does not mean a return to classical deterministic orthodoxy; on the contrary, we go beyond the certitudes associated with the traditional laws of quantum theory and emphasize the fundamental role of probabilities. In both classical and quantum physics, the basic laws now express possibilities. We need not only *laws*, but also *events* that bring an element of radical novelty to the description of nature. This novelty leads us to the "new kind of knowledge" anticipated by Maxwell. For Abraham De Moivre, one of the founders of the classical theory of probabilities, chance can neither be defined nor understood.[5] As we shall illustrate, we are now able to include probabilities in the formulation of the basic laws of physics. Once this is done, Newtonian determinism fails;

the future is no longer determined by the present, and the symmetry between past and future is broken. This confronts us with the most difficult questions of all: What are the roots of time? Did time start with the "big bang"? Or does time preexist our universe?

These questions place us at the very frontiers of space and time. A detailed explanation of the cosmological implications of our position would require a special monograph. Briefly stated, however, we believe that the big bang was an event associated with an instability within the medium that produced our universe. It marked the start of our universe but not the start of time. Although our universe has an age, the medium that produced our universe has none. Time has no beginning, and probably no end.

But here we enter the world of speculation. The main purpose of this book is to present the formulation of the laws of nature within the range of low energies. This is the domain of macroscopic physics, chemistry, and biology. It is the domain in which human existence actually takes place.

The problems of time and determinism have remained at the core of Western thought since the pre-Socratics. How can we conceive of human creativity or ethics in a deterministic world?

This question reflects a profound contradiction in Western humanistic tradition, which emphasizes the importance of knowledge and objectivity, as well as individual responsibility and freedom of choice as implied by the ideal of democracy. Popper and many other philosophers have pointed out that we are faced with an unsolvable problem as long as nature is described solely by a deterministic science.[6] Considering ourselves as distinct from the natural world would imply a dualism that is difficult for the modern mind to accept. Our aim in this work is to show that we can now overcome this obstacle. If "the passion of the western world is to reunite with the ground of its being," as Richard Tarnas has written, perhaps it is not too bold to say that we are closing in on the object of our passion.[7]

Mankind is at a turning point, the beginning of a new rationality in which science is no longer identified with certitude and probability with ignorance. We agree completely with Yvor Leclerc when he writes, "In the present century we are suffering from the separation of science and philosophy which followed upon the triumph of Newtonian physics in the eighteenth century.[8] Jacob Bronowski beautifully expressed the same thought in this way: "The understanding of human nature and of the human condition within nature is one of the central themes of science."[9]

At the end of this century, it is often asked what the future of science may be. For some, such as Stephen W. Hawking in his *Brief History of Time,* we are close to the end, the moment when we shall be able to read the "mind of God."[10] In contrast, we believe that we are actually at the beginning of a new scientific era. We are observing the birth of a science that is no longer limited to idealized and simplified situations but reflects the complexity of the real world, a science that views us and our creativity as part of a fundamental trend present at all levels of nature.

Chapter 1

EPICURUS' DILEMMA

I

Is the universe ruled by deterministic laws? What is the nature of time? These questions were formulated by the pre-Socratics at the very start of Western rationality. After more than twenty-five hundred years, they are still with us. However, recent developments in physics and mathematics associated with chaos and instability have opened up different avenues of investigation. We are beginning to see these problems, which deal with the very position of mankind in nature, in a new light, and can now avoid the contradictions of the past.

The Greek philosopher Epicurus was the first to address a fundamental dilemma. As a follower of Democritus, he believed that the world is made of atoms and the void. Moreover, he concluded, atoms fall through the void at the same speed and on parallel paths. How then could they collide? How could novelty associated with combinations of atoms ever appear? For Epicurus, the problems of sci-

9

ence, the intelligibility of nature, and human destiny could
not be separated. What could be the meaning of human
freedom in a deterministic world of atoms? As Epicurus
wrote to Meneceus, "Our will is autonomous and independent and to it we can attribute praise or disapproval.
Thus, in order to keep our freedom, it would have been
better to remain attached to the belief in gods rather than
being slaves to the fate of the physicists: The former gives
us the hope of winning the benevolence of deities through
promise and sacrifices; the latter, on the contrary, brings
with it an inviolable necessity."[1] How contemporary this
quotation sounds! Again and again, the greatest thinkers in
Western tradition, such as Immanuel Kant, Alfred North
Whitehead, and Martin Heidegger, felt that they had to
make a tragic choice between an alienating science or an
antiscientific philosophy. They attempted to find some
compromise, but none proved to be satisfactory.

Epicurus thought that he had found a solution to this
dilemma, which he termed the *clinamen*. As expressed by
Lucretius, "While the first bodies are being carried down-
wards by their own weight in straight lines through the
void, *at times quite uncertain and at uncertain places, they deviate slightly from their course,* just enough to be defined as having changed direction."[2] But no mechanism was given for
this clinamen. No wonder that it has always been considered a foreign, arbitrary element.

But do we need this novelty at all? For Heraclitus, as understood by Popper, "Truth lies in having grasped the
essential *becoming* of nature, i.e., having represented it as
implicitly infinite, as a *process in itself.*"[3] Parmenides took
the opposite view. In his celebrated poem on the unique
reality of existence, he wrote, "Nor was it ever, nor will it
be, since now it is, all together."[4]

It is amusing that the Epicurus clinamen has appeared repeatedly in the science of our century. In his classic paper on the emission of photons associated with the transitions between atomic states (1916), Einstein explicitly expressed his confidence in scientific determinism, although he assumed that these emissions are ruled by chance.

Greek philosophy was unable to solve this dilemma. Plato linked truth with being, that is, with the unchanging reality beyond becoming. Yet he was conscious of the paradoxical character of this position because it would debase both life and thought. In *The Sophist,* he concluded that we need both being and becoming.[5]

This duality has plagued Western thought ever since. As observed by the French philosopher Jean Wahl, the history of Western philosophy is, on the whole, an unhappy one, characterized by perpetual oscillations between the world as an automaton and a theology in which God governs the universe.[6] Both are forms of determinism.

This debate took a turn in the eighteenth century with the discovery of the "laws of nature." The foremost example was Newton's law relating force and acceleration, which was both deterministic and, more important, time reversible. Once we know the initial conditions, we can calculate all subsequent states as well as the preceding ones. Moreover, past and future play the same role because Newton's law is invariant with respect to the time inversion $t \rightarrow -t$. This leads to nightmares such as the demon imagined by Pierre-Simon de Laplace, capable of observing the current state of the universe and predicting its evolution.[7]

As is well known, Newton's law has been superseded in the twentieth century by quantum mechanics and relativity. Still, the basic characteristics of his laws—determinism and time symmetry—have survived. It is true that quantum mechanics no longer deals with trajectories but with wave functions (see Section IV of this chapter and Chapter 6), but it is important to note that the basic equation of quantum mechanics, Schrödinger's equation, is once again deterministic and time reversible.

By way of such equations, laws of nature lead to certitudes. Once initial conditions are given, everything is determined. Nature is an automaton, which we can control, at least in principle. Novelty, choice, and spontaneous action are real only from our human point of view.

Many historians believe that an essential role in this vision of nature was played by the Christian God as conceived in the seventeenth century as an omnipotent legislator. Theology and science agreed. As Gottfried von Leibniz wrote, "In the least of substances, eyes as piercing as those of God could read the whole course of things in the universe, *quae sint, quae fuerint, quae mox futura trahantur*" (those which are, which have been, and which shall be in the future).[8] The discovery of nature's deterministic laws was thus bringing human knowledge closer to the divine, atemporal point of view.

The concept of a passive nature subject to deterministic and time-reversible laws is quite specific to the Western world. In China and Japan, nature means "what is by itself." In his excellent book *Science and Society in East and West*, Joseph Needham tells us of the irony with which Chinese men of letters greeted the Jesuits' announcement of the triumphs of modern science.[9] For them, the idea that nature is governed by simple, knowable laws seemed to be a perfect example of anthropocentric foolishness. According to Chinese tradition, nature is spontaneous harmony; speaking about "laws of nature" would thus subject nature to some external authority.

In a message to the great Indian poet, Rabindranath Tagore, Einstein wrote:

> If the moon, in the act of completing its eternal path round the earth, were gifted with self-consciousness, it would feel thoroughly convinced that it would travel its path on its own, in accordance with a resolution taken once and for all.
>
> So would a Being, endowed with higher insight and more perfect intelligence, watching man and his doings, smile about this illusion of his that he was acting according to his own free will.
>
> This is my belief, although I know well that it is not fully demonstrable. If one thinks out to the very last consequence what one exactly knows and understands, there would hardly be any human being who could be impervious to this view, provided his self-love did not rub up against it. Man defends himself from being regarded as an impotent object in the course of the Universe. But should the lawfulness of happenings, such as unveils itself more and more clearly in inorganic nature, cease to function in the activities in our brain?[10]

To Einstein, this appeared to be the only position compatible with the achievements of science. But this conclusion is as difficult to accept now as it was to Epicurus. Time is our basic existential dimension. Since the nineteenth century, philosophy has become more and more time centered, as we see in the work of Georg Wilhelm Hegel, Edmund Husserl, William James, Henri Bergson, Martin Heidegger, and Alfred North Whitehead. For physicists such as Einstein, the problem has been solved. For philoso-

the study of being

phers, it remains the central question of ontology, at the very basis of the meaning of human existence.

In *The Open Universe:An Argument for Indeterminism,* Popper wrote, "I regard Laplacian determinism—confirmed as it may seem to be by the *prima facie* deterministic theories of physics, and by their marvelous success—as the most solid and serious obstacle to our understanding and justifying the nature of human freedom, creativity, and responsibility." For Popper, "The reality of time and change is the crux of realism."[11]

In his short essay, "The Possible and the Real," Bergson argued, "What is the role of time? . . . Time prevents everything from being given at once. . . . Is it not the vehicle of creativity and choice? Is not the existence of time the proof of indeterminism in nature?"[12] For both Popper and Bergson, we need "indeterminism." But how do we go beyond determinism? This difficulty is well analyzed in an essay by William James entitled "The Dilemma of Determinism."[13] In accord with well-defined mechanisms, determinism is "mathematizable," as shown by the laws of nature formulated by Newton, Schrödinger, and Einstein. In contrast, deviations from determinism seem to introduce anthropomorphic concepts such as chance or accident.

The conflict between the time-reversible view of physics and time-centered philosophy has led to an open clash. What is the purpose of science if it cannot incorporate some of the basic aspects of human experience? The antiscientific attitude of Heidegger is well known. Already Friedrich Nietzsche had concluded that there are no facts, only interpretations. As stated by John R. Searle, postmodern philosophy, with its idea of deconstruction, challenges Western traditions regarding the nature of truth, objectivity, and reality.[14] In addition, the role of evolution, of events, in our description of nature is steadily increasing. How then can we maintain a time-reversible view of physics?

In October 1994, there appeared a special issue of *Scientific American* devoted to "life in the universe."[15] At all levels—cosmology, geology, biology, and human society—we see a process of evolution in regard to instabilities and fluctuations. We therefore cannot avoid the question: How are these evolutionary patterns rooted in the fundamental laws of physics? Only one article, written by the eminent physicist Steven Weinberg, is relevant to this problem. He writes, "As much as we would like to take a unified view of nature, we keep encountering a stubborn duality in the role of intelligent life in the universe, as both subject and student. . . . On the one hand, there is the Schrödinger equation, which describes in a perfectly *deterministic* way how the wave function of any system changes with time. Then, quite separate, there is a set of principles that tells how to use the wave function to calculate the probabilities of various possible outcomes when someone makes the *measurement*."[16]

Does this suggest that through our measurements, we ourselves are at the origin of cosmic evolution? Weinberg speaks of a stubborn duality, a point of view found in many recent publications such as Stephen W. Hawking's *Brief History of Time*.[17] Here Hawking advocates a purely geometrical interpretation of cosmology. In short, time would be an accident of space. But he understands that this interpretation is not enough. We need an arrow of time to deal with intelligent life. Therefore, along with many other cosmologists, Hawking introduces the so-called *anthropic principle.* Nevertheless, this principle is as arbitrary as was Epicurus' clinamen. Hawking gives no indication of how the anthropic principle could ever emerge from a static geometrical universe.

As mentioned earlier, Einstein attempted to maintain the unity of nature, including mankind, at the cost of reducing us to mere automata. This was also the view of Baruch Spinoza. But there was another approach suggested by René Descartes, also in the seventeenth century, which involved the concept of dualism: on one side is matter, *res extensa,* as described by geometry, and on the other, the mind, associated with *res cogitans.*[18] In this way, Descartes described the striking difference between the behavior of simple physical systems such as a frictionless pendulum and the functioning of the human brain. Curiously, the anthropic principle brings us back to Cartesian dualism.

In *The Emperor's New Mind,* Roger Penrose states, "It is our present lack of understanding of the fundamental laws of physics that prevents us from coming to grips with the concept of 'mind' in physical or logical terms."[19] We believe that Penrose is right: We need a new formulation of the fundamental laws of physics. The evolutionary aspects of nature have to be expressed in terms of the basic laws of physics. Only in this way can we give a satisfactory answer to Epicurus' dilemma. The reasons for indeterminism, for temporal asymmetry, must be rooted in dynamics. Formulations that do not contain these features are incomplete, exactly as would be formulations of physics that ignore gravitation or electricity.

Probability plays an essential role in most sciences, from economics to genetics. Still, the idea that probability is merely a state of mind has survived. We now have to go a step farther and show how probability enters the fundamental laws of physics, whether classical or quantum. A new formulation of the laws of nature is now possible. In this way, we obtain a more acceptable description in which there is room for both the laws of nature and novelty and creativity.

At the beginning of this chapter, we mentioned the pre-Socratics. In fact, we owe to the ancient Greeks two ideals that have since shaped human history. The first is the intelligibility of nature, or in Whitehead's words, "the attempt to frame a coherent, logical, necessary system of general ideas in terms of which every element of our experience can be interpreted."[20] The second is the idea of democracy based on the assumption of human freedom, creativity, and responsibility. As long as science led to the description of nature as an automaton, these two ideals were contradictory. It is this contradiction that we are beginning to overcome.

II

In Section I, we emphasized that the problems of time and determinism form the dividing line between science and philosophy, or alternatively, between C. P. Snow's two cultures.[21] But science is far from being a monolithic bloc. In fact, the nineteenth century left us a double heritage: the laws of nature, such as Newton's law, which describes a time-reversible universe, and an evolutionary description associated with entropy.

Entropy is an essential part of thermodynamics, the science that deals specifically with irreversible, time-oriented processes. Everyone is to some extent familiar with these processes. Think about radioactive decay, or about viscosity, which slows the motion of a fluid. In contrast to time-reversible processes, such as the motion of a frictionless pendulum, where future and past play the same role (we can interchange future, that is, $+t$, with past, $-t$), irreversible processes have a direction in time. A radioactive substance prepared in the past will disappear in the future. Because of viscosity, the liquid flow slows over time.

The primordial role of the direction of time is evident in the processes we study at the macroscopic level, such as chemical reactions or transport processes. We start with chemical compounds that may react. As time goes on, they reach equilibrium and the reaction stops. Similarly, if we start with an inhomogeneous state, diffusion will tend to homogenize the system. Solar radiation is the result of irreversible nuclear processes. No description of the ecosphere would be possible without taking into account the innumerable irreversible processes that determine weather and climate. Nature involves both *time-reversible* and *time-irreversible* processes, but it is fair to say that irreversible processes are the rule and reversible processes the exception. Reversible processes correspond to idealizations: We have to ignore friction to make the pendulum move reversibly. Such idealizations are problematic because there is no absolute void in nature. As previously mentioned, time-reversible processes are described by equations of motion, which are invariant with respect to time inversion, as is the case in Newton's equation in classical mechanics or Schrödinger's equation in quantum mechanics. For irreversible processes, however, we need a description that breaks time symmetry.

+ ENTROPY

The distinction between reversible and irreversible processes was introduced through the concept of entropy associated with the so-called second law of thermodynamics. Entropy had already been defined by Rudolf Julius

Clausius in 1865 (in Greek, entropy simply means "evolution").[22] According to this law, irreversible processes produce entropy. In contrast, reversible processes leave the entropy constant.

We shall come back repeatedly to this second law. For now, let us recall Clausius's celebrated formulation: "The energy of the universe is constant. The entropy of the universe is increasing." This increase in entropy is due to the irreversible processes that take place in the universe. Clausius's statement was the first formulation of an evolutionary view of the universe based on the existence of these processes. Arthur Stanley Eddington called entropy the "arrow of time."[23] Nevertheless, according to the fundamental laws of physics, there should be no irreversible processes. We therefore see that we have inherited two conflicting views of nature from the nineteenth century: the time-reversible view based on the laws of dynamics and the evolutionary view based on entropy. How can these conflicting views be reconciled? After so many years, this problem is still with us.

For the Viennese physicist Ludwig Boltzmann, the nineteenth century was the century of Charles Darwin, the man who defined life as the result of a never-ending process of evolution and thus placed becoming at the center of our understanding of nature. Still, for most physicists, Boltzmann is now associated with a conclusion quite opposite to that of Darwin; he is credited with having shown that irreversibility is only an illusion. It was Boltzmann's tragedy to have attempted in physics what Darwin had accomplished in biology—only to reach an impasse.

At first glance, the similarities between the approaches of these two giants of the nineteenth century are striking. Darwin showed that if we start with the study of populations, and not individuals, we can understand how individual variability, subject to selection pressure, produces a drift. Correspondingly, Boltzmann argued that we cannot understand the second law of thermodynamics, and the spontaneous increase in entropy it predicts, by starting with individual dynamical trajectories; we must begin instead with large populations of particles. The increase in entropy would be the global drift resulting from the numerous collisions between these particles.

In 1872, Boltzmann published his famous H-theorem, which included the H-function, a microscopic analogue of entropy.[24] This theorem takes into account the effects of collisions that modify the velocities of particles at each instant. It shows that collisions bring the distribution of velocities of the population of particles closer to equilibrium (the so-called Maxwell-Boltzmann distribution). As the population approaches equilibrium, Boltzmann's H-function decreases and reaches its minimum value at equilibrium; this minimum value means that collisions no longer modify the distribution of velocities. For Boltzmann, the particle collisions are thus the mechanism that leads the system to equilibrium.

Both Boltzmann and Darwin replaced the study of "individuals" with the study of populations, and showed that slight variations (the variability of individuals, or microscopic collisions) taking place over a long period of time can generate evolution at a collective level. (In later chapters, we shall come back to the role of populations.) Exactly as biological evolution cannot be defined at the level of individuals, the flow of time is also a global property (see Chapters 5 and 6). But while Darwin attempted to explain the appearance of new species, Boltzmann described an evolution toward equilibrium and uniformity. Significantly, these two theories have had very different fortunes. Darwin's theory of evolution, which was to triumph in spite of fierce controversies, remains the basis for our understanding of life. On the other hand, Boltzmann's interpretation of irreversibility succumbed to its critics, and he was gradually forced to retreat. He could not exclude the possibility of "antithermodynamical" evolutions, as a result of which entropy would diminish and inhomogeneities, instead of being leveled, would increase spontaneously.

The situation confronting Boltzmann was indeed dramatic. He was convinced that in order to understand nature we have to include evolutionary features and that irreversibility, as defined by the second law of thermodynamics, was a decisive step in this direction. But he was also heir to the grand tradition of dynamics, and realized that it stood in the way of his attempt to give a microscopic meaning to the arrow of time.

From today's vantage point, Boltzmann's need to choose between his conviction that physics had to understand becoming, and his loyalty to its traditional role, seems particularly poignant. The fact that his attempt would end in failure now seems self-evident. Every student learns that a trajectory is time reversible, and thus allows no distinction between future and past. As Henri Poincaré noted, explaining irreversibility in terms of trajectories that are time-reversible processes, however numerous, appears to be a purely logical error.[25] Suppose that we invert the sign of the velocity of all molecules. The system would then go into its own "past." Even if entropy was increasing before velocity inversion, it would now decrease. This was Joseph Loschmidt's velocity-reversal paradox, which was the reason why Boltzmann could not exclude antithermodynamical behavior. When faced with severe criticism, he re-placed his microscopic interpretation of the second law with a *probabilistic* interpretation based upon our lack of in-formation.

In a complex system formed by huge numbers of mol-ecules (on the order of 10^{23}, or Avogadro's number), such as a gas or liquid, it is obvious that we are unable to com-pute the behavior of each molecule. For this reason, Boltz-mann introduced the assumption that all microscopic states of such a system have the same *a priori* probability. The dif-ference would be associated with the macroscopic state, as described by temperature, pressure, and other parameters. Boltzmann defined the probability of each macroscopic state by calculating the number of microscopic states that give rise to it.

Boltzmann would have us imagine, for instance, a vol-ume divided into two equal compartments that communi-cate with each other. This volume contains a large number of molecules, which we shall call N. Although we are un-able to follow the path of each individual molecule, through measuring a macroscopic quantity, such as the pressure in each compartment, we can determine the number of molecules it contains. We can also prepare a starting point, or "initial state" as it is generally referred to by physicists, where one of the two compartments is nearly empty. What can we expect to observe? Over the course of time, molecules will populate the empty com-partment. Indeed, the great majority of all possible micro-scopic states corresponds to a macroscopic situation where each compartment contains the same number of mole-cules. These states correspond to equilibrium, or to pres-sures that would be equal in the two compartments. Once this state has been achieved, the molecules will continue to pass from one compartment to the other, but on average, the number of molecules going to the right and left will be equal. Apart from slight transitory fluctuations, the number of molecules in the two compartments will remain constant over time, and equilibrium will be preserved. However, there is a basic weakness in this argument. A spontaneous long-term deviation from equilibrium is not impossible, even if it is, as Boltzmann concluded, "improbable."

Boltzmann's probability-based interpretation makes the macroscopic character of our observations responsible for the irreversibility we observe. If we could follow the individual motion of the molecules, we would see a time-reversible system in which each molecule follows the laws of Newtonian physics. Because we can only describe the number of molecules in each compartment, we conclude that the system evolves toward equilibrium. According to this interpretation, irreversibility is not a basic law of nature; it is merely a consequence of the approximate, macroscopic character of our observations.

Ernst Zermelo added another criticism of Boltzmann's argument to Loschmidt's reversal paradox[26] in quoting Poincaré's recurrence theorem, which shows that if we were to wait long enough, we could observe the spontaneous return of a dynamical system to a state as close to the initial state as we might wish. As the physicist Roman Smoluchowski concluded, "If we continued our observation for an immeasurably long time, all processes would appear to be reversible."[27] This applies directly to Boltzmann's two-compartment model. After a sufficiently long time, the initially empty compartment will again become empty. Irreversibility corresponds only to an appearance that is devoid of any fundamental significance.

Let us now return to the situation discussed in Section I. Through our own approximations, we would be responsible for the evolutionary character of the universe. In order to make such an argument plausible, the first step in assuring that irreversibility will be the result of our approximations is to view the consequences of the second law as trivial and self-evident. In his recent book, *The Quark and the Jaguar*, Murray Gell-Mann writes,

The explanation [of irreversibility] is that there are more ways for nails or pennies to be mixed up than sorted. There are more ways for peanut butter and jelly to contaminate each other's containers than there are to remain completely pure. And there are more ways for oxygen and nitrogen gas molecules to be mixed up than segregated. To the extent that chance is operating, it is likely that a closed system that has some order will move toward disorder, which offers so many more possibilities. How are those possibilities to be counted? An entire closed system, exactly described, can exist in a variety of states, often called microstates. In quantum mechanics, these are understood to be possible quantum states of the system. These microstates are grouped into categories (sometimes called macrostates) according to the various properties that are being distinguished by *coarse graining*. The microstates in a given macrostate are then treated as equivalent, so that only their number matters. . . .

Entropy and information are very closely related. In fact, entropy can be regarded as a measure of ignorance. When it is known only that a system is in a given macrostate, the entropy of the macrostate measures the degree of ignorance the microstate system is in by counting the number of bits of additional information needed to specify it, with all the microstates in the macrostate treated as equally probable.[28]

Similar arguments can be found in most books dealing with the arrow of time. We believe that these arguments are untenable. They imply that it is our own ignorance, our coarse graining, that leads to the second law. For a well-informed observer, such as the demon imagined by Laplace, the world would appear as perfectly time reversible. We would be the father of time, of evolution, and not its children. Irreversibility subsists, whatever the precision of our experiments. This means that attributing these properties to incomplete information can hardly be considered. It is interesting to note that Max Planck had already opposed the idea of incomplete information to describe the second law. In his *Treatise on Thermodynamics* he wrote,

It would be absurd to assume that the validity of the second law depends in any way on the skill of the physicist or chemist in observing or experimenting. The gist of the second law has nothing to do with experiment; the law asserts briefly that *there exists in nature a quantity which always changes in the same way in all natural processes.* The proposition stated in this general form may be correct or incorrect; but whichever it may be, it will remain so, irrespective of whether thinking and measuring beings exist on the earth or not, and whether or not, assuming they do exist, they are able to measure the details of physical or chemical processes more accurately by one, two, or a hundred decimal places than we can. The limitation of the law, if any, must lie in the same province as its essential idea, in the observed Nature, and not in the Observer. That man's experience is called upon in the deduction of the law is of no consequence; for that is, in fact, our only way of arriving at a knowledge of natural law.[29]

However, Planck's views were to remain isolated. As we have indicated, most scientists considered the second law the result of approximations, or the intrusion of subjective views into the exact world of physics. In a celebrated statement, Max Born asserted that "irreversibility is the effect of the introduction of ignorance into the basic laws of physics."[30]

Our own point of view is that the laws of physics, as formulated in the traditional way, describe an idealized, stable world that is quite different from the unstable, evolving world in which we live. The main reason to discard the banalization of irreversibility is that we can no longer associate the arrow of time only with an increase in disorder. Recent developments in nonequilibrium physics and chemistry point in the opposite direction. They show unambiguously that the arrow of time is a source of *order*. This is already clear in simple experiments such as thermal diffusion, which has been known since the nineteenth century. Let us consider a box containing two components (such as hydrogen and nitrogen) where we heat one boundary and cool the other (see Figure 1.1). The system evolves to a steady state in which one component is enriched in the hot part and the other in the cold part. The entropy produced by the irreversible heat flow leads to an ordering process, which would be impossible if taken independently from the heat flow. Irreversibility leads to both order and disorder.

The constructive role of irreversibility is even more striking in far-from-equilibrium situations where nonequilibrium leads to new forms of coherence. (We shall come back to nonequilibrium physics in Chapter 2.) We have now learned that it is precisely through irreversible processes associated with the arrow of time that nature achieves its most delicate and complex structures. Life is

Figure 1.1
Thermal Diffusion

As a result of the difference in temperature between the two containers, the black molecules have a higher concentration in the left compartment. This corresponds to thermal diffusion.

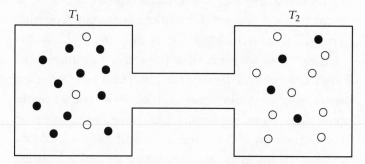

possible only in a nonequilibrium universe. Nonequilibrium leads to concepts such as self-organization and dissipative structures, which will be described in more detail in Chapter 2. In *From Being to Becoming*, we had already formulated the following conclusions based on the remarkable developments in nonequilibrium physics and chemistry over the past several decades:

- Irreversible processes (associated with the arrow of time) are as real as reversible processes described by the fundamental laws of physics; they do not correspond to approximations added to the basic laws.
- Irreversible processes play a fundamental constructive role in nature.[31]

What impact do these concepts have on current thinking about dynamical systems? Boltzmann was well aware that nothing analogous to irreversibility exists in classical dynamics; he therefore concluded that irreversibility can be derived only from assumptions about the initial conditions in the early stages of our universe. We can keep our usual formulations of dynamics, but we need to supplement them with appropriate initial conditions. In this view, the original universe was highly organized, and therefore in an improbable state—a suggestion still accepted in a number of recent books.[32] The initial conditions prevailing in our universe lead to interesting and largely unsolved problems (see Chapter 8), but we believe that Boltzmann's argument is no longer defensible. Whatever the past, there exist at present two types of processes: time-reversible processes, where the application of existing dynamics has proved to be successful (i.e., the motion of the moon in classical mechanics, or the hydrogen atom in quantum mechanics), and irreversible processes like heat conditions, where the asymmetry between past and future is obvious. Our objective is to devise a new formulation of physics that explains, independently of any cosmological considerations, the difference between these behaviors. This can indeed be achieved for unstable and thermodynamic systems. We can overcome what looked like an apparent contradiction between the time-reversible laws of dynamics and the evolutionary view of nature based on entropy. But let us not get ahead of ourselves.

Nearly two hundred years ago, Joseph-Louis Lagrange described analytical mechanics based on Newton's laws as a branch of mathematics.[33] In the French scientific literature, one often speaks of "rational mechanics." In this sense, Newton's laws would define the laws of reason and represent a truth of absolute generality. Since the birth of quantum mechanics and relativity, we know that this is not the case. The temptation is now strong to ascribe a similar status of absolute truth to quantum theory. In *The Quark and the Jaguar,* Gell-Mann asserts, "Quantum mechanics is

PAST

not itself a theory; rather it is the framework into which all contemporary physical theory must fit."[34] Is this really so? As stated by my late friend Léon Rosenfeld, "Every theory is based on physical concepts expressed through mathematical idealizations. They are introduced to give an adequate representation of the physical phenomena. *No physical concept is sufficiently defined without the knowledge of its domain of validity.*"[35]

It is this "domain of validity" required for the basic concepts of physics, such as trajectories in classical mechanics or wave functions in quantum theory, that we are beginning to delineate. These limits are associated with instability and chaos, which we shall briefly introduce in the next section. Once we include these concepts, we come to a new formulation of the laws of nature, one that is no longer built on certitudes, as is the case for deterministic laws, but rather on *possibilities.* Moreover, in this probabilistic formulation, time symmetry is destroyed. The evolutionary character of the universe has to be reflected within the context of the fundamental laws of physics. Remember the ideal of the intelligibility of nature as formulated by Whitehead (see Section I): Every element of our experience has to be included in a coherent system of general ideas. Based on this rewriting of the laws of nature, we can now complete the work pioneered by Boltzmann more than a century ago.

It is interesting that great mathematicians, such as Emile Borel, also understood the need to overcome determinism. Borel noted that considerations of isolated systems, such as moon-earth, are always idealizations, and that determinism may fail when we leave this reductionist view.[36] That is indeed what our own research shows.

III

Everyone is to some extent familiar with the difference between stable and unstable systems. Consider a pendulum, for example. Suppose that it is originally at equilibrium, where the potential energy is at a minimum. If a small perturbation is followed by a return to equilibrium (see Figure 1.2), this system represents a *stable* equilibrium. In contrast, if we put a pencil on its head, the smallest perturbation will cause it to fall to the left or right, giving us a model of *unstable* equilibrium.

There is a basic distinction between stable and unstable motions. In short, stable dynamical systems are those in which slight changes in the initial conditions produce correspondingly slight effects. But for a large class of dynamical systems, small perturbations in the initial conditions are amplified over the course of time. Chaotic systems are an extreme example of unstable motion because trajectories identified by distinct initial conditions, no matter how close, diverge exponentially over time. This is known as "sensitivity to initial conditions." A classic illustration of amplification through chaos is the "butterfly effect"; by

Figure 1.2
Stable and Unstable Equilibrium

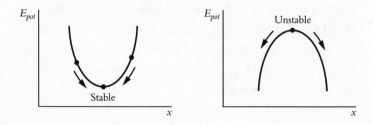

just flapping its wings, a butterfly in Amazonia may affect the weather in the United States. We shall see examples of chaotic systems later on in Chapters 3 and 4.

The term *deterministic chaos* has also entered the discussion of chaotic systems. Indeed, the equations of motion remain deterministic, as is the case in Newtonian dynamics, even if a particular outcome appears to be random. The discovery of the important role of instability has led to a revival of classical dynamics, previously considered a closed subject. In fact, until recently it was thought that all systems described by Newton's laws are alike. Of course, everyone knew that the trajectory of a falling stone was easier to solve than a "three-body problem," such as the one involving the sun, Earth, and Jupiter. But this was considered to be merely a question of computation. It was only at the end of the nineteenth century that Poincaré showed that this was not the case. The problems are fundamentally different depending on whether or not a dynamical system is stable.

We have mentioned chaotic systems, but there are other types of instability to be considered. Let us first describe in qualitative terms in what sense instability leads to an extension of the laws of dynamics. In classical dynamics, the initial state is determined by the positions q and velocities v (or momentum p).* Once these are known, we can determine the trajectory by using Newton's laws (or any other equivalent formulation of dynamics). We can then represent the dynamical state by a point q_0, p_0 in a space formed by the coordinates and momenta. This is known as the *phase space* (Figure 1.3). Instead of examining a single system, we can also study a collection of systems—an "en-

---

*For the purpose of simplification, we have used a single letter even when we are discussing a system formed by many particles.

Figure 1.3

Trajectory in Phase Space

The dynamical state is represented by a point in the phase space q, p. The time evolution is described by a trajectory starting at the initial point q_0, p_0.

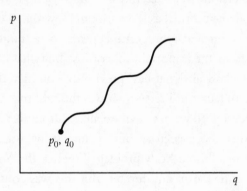

semble," as it has been called since the pioneering work of Albert Einstein and Josiah Willard Gibbs at the beginning of this century.

At this point, it would be helpful to reproduce part of Gibbs's famous preface to his *Elementary Principles in Statistical Mechanics*:

> We may imagine a great number of systems of the same nature, but differing in the configurations and velocities which they have at a given instant, and differing not merely infinitesimally, but it may be so as to embrace every conceivable combination of configurations and velocities. And here we may set the problem, not to follow a particular system through its succession of configurations, but to determine how the whole number of systems will be distributed among the various conceivable configurations and velocities at any required time, when the distribution has been given for some one time. . . .

The laws of thermodynamics, as empirically determined, express the *approximate and probable* behavior of systems of a great number of particles, or, more precisely, they express the laws of mechanics for such systems as they appear to beings who have not the fineness of perception to appreciate quantities of the order of magnitude of those which relate to single particles, and who cannot repeat their experiments often enough to obtain any but the most probable results.[37]

Gibbs introduced population dynamics into physics by using an ensemble approach. An ensemble is represented by a cloud of points in phase space (see Figure 1.4). The cloud is described by a function $\rho(q,p,t)$, which has a simple physical interpretation: the *probability* of finding at time t, a point in the small region of phase space around the point q,p. A trajectory corresponds to a special case in which ρ is vanishing everywhere except at the point q_0,p_0. This situation is described by a special form of ρ. Functions that have the property of vanishing everywhere except at a single point are called Dirac delta functions $\delta(x)$. The function $\delta(x - x_0)$ is vanishing for all points $x \neq x_0$. Therefore, for a single trajectory at time zero, the distribution function ρ takes the form $\rho = \delta(q - q_0)\delta(p - p_0)$.* We shall come back to the properties of delta functions later.

---

*When we take $x = x_0$, the function $\delta(x - x_0)$ diverges to infinity. The δ-function therefore has "abnormal" properties as compared to a continuous function such as x or $\sin x$. It is called a *generalized function* or distribution (not to be confused with probability distribution ρ). Generalized functions are used in conjunction with test functions $\varphi(x)$, which are continuous functions (i.e., $\int dx \varphi(x)\delta(x - x_0) = \varphi(x_0)$). Also note that at time t we have for a free particle moving at the speed $\frac{p_0}{m}$ the probability $\rho = \delta(p - p_0)\delta(q - q_0 - \frac{p_0 t}{m})$, as the momentum remains constant and the coordinate varies linearly with time.

Figure 1.4

Ensembles in Phase Space

Gibbs's ensemble is represented by a cloud of particles differing according to their initial conditions. The shape of the cloud changes over time.

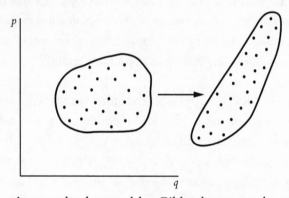

As was clearly stated by Gibbs, however, the ensemble approach was merely a convenient computational tool for him when exact initial conditions were not available. In his opinion, probabilities express ignorance, or lack of information. Moreover, it has always been accepted that from the dynamical point of view, individual trajectories and probability distributions present equivalent problems. We can start with individual trajectories and then derive the evolution of probability functions, and vice versa. The probability ρ corresponds simply to a superposition of trajectories, and leads to no new properties. The two levels of description, the *individual* level (corresponding to single trajectories) and the *statistical* level (corresponding to ensembles), would be equivalent.

Is this always the case? For simple stable systems, where we do not expect any irreversibility, this is indeed true. Gibbs and Einstein were right. The individual point of view (in terms of trajectories) and the statistical point of view (in terms of probabilities) are then equivalent. This can be easily verified, and we shall come back to this point in Chapter 5. However, is this also true for unstable systems? How is it that all theories dealing with irreversible processes on the molecular level, such as Boltzmann's kinetic theory, involve probabilities and not trajectories? Is this again because of our approximations, our coarse graining? How can we then explain the success of kinetic theory, the quantitative predictions of many properties of dilute gases, such as thermal conductivity and diffusion, all of which have been verified by experimentation?

Poincaré was so impressed by the success of kinetic theory that he wrote, "Perhaps the kinetic theory of gases will serve as a model . . . Physical laws will then take on a completely new form; they will take on a *statistical character.*"[38] These were indeed prophetic words. In an extraordinarily daring move, Boltzmann introduced probability as an empirical tool. Now, more than one hundred years later, we are beginning to understand how probabilistic concepts emerge when we go from dynamics to thermodynamics. Instability destroys the equivalence between the individual and statistical levels of description. Probabilities then acquire an intrinsical dynamical meaning. This knowledge has led to a new kind of physics, the physics of populations, which is the basic subject of this book.

By way of explanation, let us consider a simplified example of chaos. Suppose that we have two types of motion denoted as + or − (i.e., motion "up" or "down") within the phase space illustrated in Figure 1.4. This leads us to the two types of situations represented by Figures 1.5 and 1.6. In Figure 1.5, there are two different regions in phase space, one corresponding to the motion −, the other to the motion +. If we discard the region close to the boundary, each − is surrounded by −, and each + by +. This corre-

Figure 1.5
Stable Dynamical System
The motions denoted as + or − lie in distinct regions of phase space.

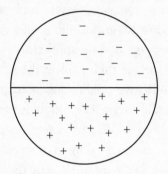

sponds to a stable system. Slight changes in the initial conditions do not alter the result.

In Figure 1.6, instead, each + is surrounded by −, and vice versa. The slightest change in initial conditions is amplified, and the system is therefore unstable. A primary re-

Figure 1.6
Unstable Dynamical System
Each motion + is surrounded by − and vice versa.

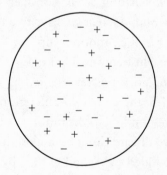

sult of this instability is that trajectories now become *idealizations*. We can no longer prepare a single trajectory, as this would imply infinite precision. For stable systems, this is without significance, but for unstable systems, with their sensitivity to initial conditions, we can only prepare probability distributions, including various types of motion.

Is this difficulty merely a practical one? Yes, if we consider that trajectories have now become uncomputable. But there is more: Probability distribution permits us to incorporate within the framework of the dynamical description the complex microstructure of the phase space. It therefore contains *additional* information that is lacking at the level of individual trajectories. As we shall see in Chapter 4, this has fundamental consequences. At the level of distribution functions ρ, we obtain a new dynamical description that permits us to predict the future evolution of the ensemble, including characteristic time scales. This is impossible at the level of individual trajectories. The equivalence between the individual and statistical levels is indeed broken. We obtain new solutions for the probability distribution ρ that are *irreducible* because they do not apply to single trajectories. The laws of chaos have to be formulated at the statistical level. That is what we meant in the preceding section when we spoke about a generalization of dynamics that cannot be expressed in terms of trajectories. This leads to a situation that has never been encountered in the past. The initial condition is no longer a point in the phase space but some region described by ρ at the initial time $t =$ zero. We thus have a *nonlocal* description. There are still trajectories, but they are the outcome of a stochastic, probabilistic process. No matter how precisely matched our initial conditions are, we obtain different trajectories from them. Moreover, as we shall see, time symmetry is broken, as past and future play different roles in the statistical formulation. Of course, for stable systems, we revert to the usual description in terms of deterministic trajectories.

Why has it taken so long to arrive at a generalization of the laws of nature that includes irreversibility and probability? One of the reasons is ideological—the desire to achieve a quasi-divine point of view in our description of nature. But there has also been a technical, mathematical problem involved. Our work is based on recent progress in functional analysis, a field of mathematics that has come to the forefront only in recent decades. As we shall see, our formulation requires an extended functional space. This new field of mathematics, which uses generalized functions or fractals, as Benoît Mandelbrot called them, is now playing a critical role in the understanding of the laws of nature.[39] We need a "divine" point of view to retain the idea of determinism. But no human measurements, no theoretical predictions, can give us initial conditions with infinite precision.

It is interesting to contemplate what becomes of the Laplace demon in the world of deterministic chaos. He can no longer predict the future unless he knows the initial conditions with infinite precision. Only then can he continue to use a trajectory description. But there is an even more powerful instability that leads to the destruction of trajectories, *whatever the precision of the initial description.* This form of instability is of fundamental importance because it applies to both classical and quantum mechanics.

Our story actually begins at the end of the nineteenth century with the work of Jules-Henri Poincaré. According to Poincaré, a dynamical system is characterized in terms of the kinetic energy of its particles plus the potential energy due to their interaction.[40] A simple example would be free, noninteracting particles, where there is no potential energy, and the calculation of trajectories is trivial. Such systems are by definition integrable. Poincaré then asked the question: Are all systems integrable? Can we choose suitable variables to eliminate potential energy? By showing that this was generally impossible, he proved that dynamical systems were largely *nonintegrable*.

It is worthwhile to pause for a moment and reflect on Poincaré's conclusions. Suppose he had proved that all dynamical systems are integrable. This would mean that all dynamical motions are isomorphic to free noninteracting particles. There would be no place for the arrow of time, for self-organization, or life itself. Integrable systems describe a static, deterministic world. Poincaré not only demonstrated nonintegrability, but also identified the reason for it: *the existence of resonances between the degrees of freedom*. As we shall see in greater detail in Chapter 5, there is a frequency that corresponds to each mode of motion. The simplest example of this is the harmonic oscillator, in which a particle and central point are given. The particle is held by a force proportional to its distance from that point. If we displace the particle from the center, it will oscillate with a well-defined frequency. It is through these frequencies that we arrive at the notion of resonance, which is crucial to Poincaré's theorem.

We are all more or less familiar with the concept of resonance. When we force a spring to deviate from its equilibrium position, it vibrates with a characteristic frequency. Now let us subject this spring to an external force with a frequency that can be varied. When the two frequencies, that of the spring and that of the external force, have a simple numerical ratio (that is, when one of the frequencies is either equal to the other, or two, three, four . . . times larger), the amplitude of the motion of the spring increases dramatically. The same phenomenon occurs when we play a note on a musical instrument. We hear the harmonics. Resonance "couples" sounds.

Now consider the case of a system characterized by two frequencies. By definition, whenever the sum $n_1\omega_1 + n_2\omega_2 = 0$, where n_1 and n_2 are nonvanishing integers, we have resonance. This means that $\frac{\omega_1}{\omega_2} = -\frac{n_2}{n_1}$; the ratio of the frequencies is then a rational number. As Poincaré has shown, in dynamics resonances lead to terms with "dangerous" denominators such as $\frac{1}{n_1\omega_1 + n_2\omega_2}$. Whenever there are resonances (i.e., points in phase space where $n_1\omega_1 + n_2\omega_2 =$ zero), these terms diverge. As a result, we encounter obstacles whenever we try to calculate trajectories.

This is the origin of Poincaré's nonintegrability. The "problem of small denominators" was already known by eighteenth-century astronomers, but Poincaré's theorem showed that this difficulty, which he called the "general problem of dynamics," is shared by the great majority of dynamical systems. For a considerable length of time, however, the importance of Poincaré's findings was overlooked.

Max Born wrote, "It would indeed be remarkable if Nature had fortified herself against further advances in knowledge behind the analytical difficulties of the many-body problem."[41] It was hard to believe that a technical difficulty (divergences due to resonances) could alter the conceptual structure of dynamics. We now see this problem in a different way. For us, Poincaré's divergences are an opportunity. Indeed, we can go beyond his negative statement and show that nonintegrability paves the way, as does chaos, for a new *statistical* formulation of the laws of dynamics. It took sixty years after Poincaré, through the work of Andrei N. Kolmogorov, continued by Vladimir Igorevich Arnold and Jürgen Kurt Moser (the so-called KAM theory), for nonintegrability to be understood not as the frustrating manifestation of some resistance of nature against the advances of knowledge, to paraphrase Born, but as a new starting point for dynamics.[42]

The KAM theory deals with the influence of resonances on trajectories. The frequencies ω depend in general on the values of dynamic variables such as coordinates and momenta. They therefore take on different values at different points in the phase space. The result is that some points will be characterized by resonances, and others will not. Again, for chaos this leads to an extraordinary complexity in the phase space. According to the KAM theory, we observe two types of trajectories: "nice" deterministic trajectories and "random" trajectories associated with resonances, which wander erratically through regions of phase space.

Another important result of this theory is that when we increase the value of energy, we increase the regions where randomness prevails. For some critical value of energy, chaos appears: over time we observe the exponential divergence of neighboring trajectories. Furthermore, for fully developed chaos, the cloud of points generated by a trajectory leads to diffusion. But diffusion is associated with the approach to uniformity in our *future*. It is an irreversible process that creates entropy (see Section I). Although we started with classical dynamics, we can now observe the breaking of time symmetry. How this is possible is the main problem we have to solve in order to overcome the time paradox.

Poincaré resonances play a fundamental role in physics. Emission or absorption of light is due to resonances, as is the approach to equilibrium in a system of interacting particles. Interacting fields again lead to resonances. It is difficult to identify an important problem in classical or quantum physics where resonances do not play a significant part. But how can we overcome the divergences associated with resonances? Here some essential progress has been made. As in Section III, we have to distinguish the individual level (trajectories) from the statistical level (ensembles, as described by the probability distribution ρ). At the individual level we have divergences, but these can be solved at a statistical level (see Chapters 5 and 6), where resonances produce a coupling of events loosely analogous to the coupling of sounds by resonance. This leads to new, non-Newtonian terms that are *incompatible with a trajectory description* and instead require a statistical, probabilistic description. This is not astonishing. Resonances are not local events, inasmuch as they do not occur at a given point or instant. They imply a nonlocal description and therefore cannot be included in the trajectory description associated with Newtonian dynamics. As we shall see, they lead to *diffusive* motion. When we start at a point P_0 in phase space, we can no longer predict with certainty its position P_τ after a time period τ. In short, the initial point P_0 leads to many possible points—P_1, P_2, P_3—with well-defined probabilities.

In Figure 1.7, each point in the domain D has a nonvanishing or well-defined transition probability of appearing at time τ. This situation is similar to that of the "random walk," or "Brownian motion." In the simplest case, this condition may be illustrated by a particle on a one-dimensional lattice that makes a one-step transition at regular time intervals (see Figure 1.8).

Figure 1.7
Diffusive Motion

After a time *t*, the system may produce a result at any point, such as P_1, P_2, P_3, in the domain *D*.

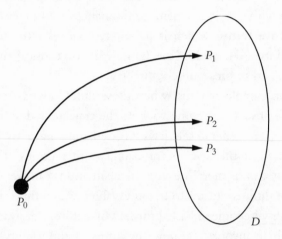

At every step, the probability is $\frac{1}{2}$ that the particle will go to the left and $\frac{1}{2}$ that it will go to the right. At every step, the future is uncertain. From the very beginning, it is impossible to speak of trajectories. Mathematically, Brownian motion is described by diffusion-type equations (the so-called Fokker-Planck equations). Since diffusion is

Figure 1.8
A Random Walk

Brownian motion on a one-dimensional lattice. At every step, the probability is $\frac{1}{2}$ that the particle will go to the left and $\frac{1}{2}$ that it will go to the right.

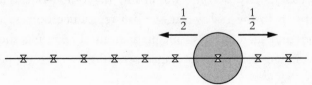

time oriented, if we start with a cloud of points, all of which are situated at the same origin, as time goes on the cloud will disperse. Some particles will be found farther from the origin, others closer. It is quite remarkable that, starting with classical dynamics, resonances lead precisely to diffusive terms, which is to say, resonances introduce uncertainty even within the framework of classical mechanics, and so break time symmetry.

For integrable systems, when these diffusive contributions are absent, we come back to the trajectory description, but in general the laws of dynamics have to be formulated at the level of probability distributions. The basic question is therefore: In which situations can we expect the diffusive terms to be observable? When this is so, probability becomes a basic property of nature. This question, which involves defining the limits of the validity of Newtonian dynamics (or the validity of quantum theory, which we shall consider in the next section), is nothing short of revolutionary. For centuries, trajectories have been considered the basic, primitive object of classical physics. In contrast, we now consider them of limited validity for resonant systems. We shall return repeatedly to this question in Chapter 5, and to a parallel question for quantum mechanics in Chapter 6. For the moment, however, let us present some provisional answers. For *transient* interactions (a beam of particles collides with an obstacle and escapes), diffusive terms are negligible. But for *persistent* interactions (a steady flow of particles falls onto the obstacle), they become dominant. In computer simulations, as in the real world, we can reproduce both situations and therefore test our predictions. The results show unambiguously the appearance of diffusive terms for persistent interactions, and therefore the breakdown of the Newtonian, as well as the orthodox, quantum mechanical descriptions. In both these cases, we obtain "irreducible" probabilistic descriptions, as in deterministic chaos.

But there is yet another situation that is even more remarkable. Macroscopic systems are generally defined in terms of the *thermodynamic limit*, according to which both the number N of particles and the volume V become large. We shall study this limit in Chapters 5 and 6. In the observation of phenomena associated with this limit, the new properties of matter become obvious.

As long as we consider merely a few particles, we cannot say if they form a liquid or gas. States of matter as well as phase transitions are ultimately defined by the thermodynamic limit. The existence of phase transitions shows that we have to be careful when we adopt a reductionist attitude. Phase transitions correspond to emerging properties. They are meaningful only at the level of populations, and not of single particles. This contention is somewhat analogous to that which is based on Poincaré resonances. Persistent interactions mean that we cannot take a part of the system and consider it in isolation. It is at this global level, at the level of populations, that the symmetry between past and future is broken, and science can recognize the flow of time. This solves a long-standing puzzle. It is indeed in macroscopic physics that irreversibility and probability are the most conspicuous.

Thermodynamics applies to non-integrable systems. This means that we cannot solve the dynamical problem in terms of trajectories, but we can solve it in terms of probabilities. Therefore, as is the case for deterministic chaos, the new statistical formulation of classical mechanics leads to an extension of the mathematical framework. To some extent, this is reminiscent of general relativity. As Einstein showed, we have to move from Euclidean geometry to Riemannian geometry to include gravitation. In functional calculus, a special role is played by the so-called Hilbert space, which extends Euclidean geometry to situations involving an infinite number of dimensions (the "function space"). Traditionally, quantum mechanics and statistical mechanics have utilized Hilbert space. To obtain our new formulation, which is valid for unstable systems and the thermodynamic limit, we have to move from Hilbert space to more general functional spaces. This observation will be explained in detail in Chapters 4 through 6.

Since the beginning of this century, we have become used to the idea that classical mechanics has to be extended when we consider microscopic objects, such as atoms or elementary particles, or when we deal with astrophysical dimensions. Surprisingly, instability also requires an extension of classical mechanics. The situation in quantum mechanics, to which we now turn, is quite similar. Instability driven by resonances plays a fundamental role in changing the formulation of quantum theory.

IV

In quantum mechanics, we encounter a rather strange situation. As is well known, this theory has been remarkably successful in all its predictions. Still, more than sixty years after its formulation, discussions about its meaning and scope are as heated as ever. This is unique in the history of science.[43] In spite of all its successes, most physicists share some feeling of uneasiness. Richard Feynman once re-marked that nobody really "understands" quantum theory. Here, the basic quantity is the wave function Ψ, which plays somewhat the role of the trajectory in classical me-chanics. Indeed, the fundamental equation of quantum theory, the Schrödinger equation, describes the time evo-lution of the wave function. It transforms the wave func-tion $\Psi(t_0)$, as given at the initial time t_0, into the wave function $\Psi(t)$ at time t, exactly as trajectories in classical mechanics lead from one phase point to another.

Like Newton's equation, Schrödinger's equation is de-terministic and time reversible. Again, as in classical dynamics, there appears a gap between the dynamical de-scription of quantum mechanics and the evolutionary description associated with entropy. The physical interpre-tation of the wave function Ψ is that of a *probability ampli-tude*. This implies that the square $|\Psi|^2 = \Psi\Psi*$ (Ψ has both a real and imaginary part; $\Psi*$ is the complex conju-gate of Ψ) is a probability, which we shall again denote by ρ. There are more general forms of probability corre-sponding to ensembles obtained by the superimposing of various wave functions. These are called mixtures, as op-posed to pure cases that obtain from a single wave func-tion.

The basic assumption of quantum theory is that every dynamical problem can be solved at the level of probabil-ity *amplitudes* exactly as every dynamical problem in classi-cal mechanics was traditionally associated with trajectory dynamics. But strangely, in order to attribute well-defined properties to matter, we have to go beyond probability am-plitudes; we need probabilities themselves. To understand this difficulty, let us consider a simple example. Suppose that energy can take on two values, E_1 and E_2. The corresponding wave function is u_1 or u_2. Now consider the linear superimposition $\Psi = c_1 u_1 + c_2 u_2$. The wave function then "participates" at both levels. The system is neither at level 1 nor level 2, but rather in a kind of intermediate state. Let us now measure the energy associated with Ψ. According to quantum mechanics, we then find either E_1 or E_2 with probabilities given by the squares of the probability amplitudes $|c_1|^2$ and $|c_2|^2$.

Initially we started with a single wave function Ψ, but we still end up with a mixture of two wave functions, u_1 and u_2. This is often called the "reduction," or "collapse," of the wave function. We need to move from *potentialities* described by the wave function Ψ to *actualities* that we can measure. In the traditional language of quantum theory, we move from a pure state (the wave function) to an ensemble, or mixture. But how is this possible? As mentioned earlier, Schrödinger's equation transforms a wave function into another wave function, and not into an ensemble. This has often been called the *quantum paradox*. It has been suggested that the transition from potentiality to actuality is due to our own measurements. This is the point of view expressed by Steven Weinberg in Section I of this chapter and in a considerable number of textbooks. It is the same type of explanation as was presented for the time paradox in classical mechanics. In that case as well, it is difficult to understand how a human action, such as observation, could be made responsible for the transition from potentialities to actualities. Would the evolution of the universe be different in the absence of humankind? In his Introduction to *The New Physics: A Synthesis,* Paul C.W. Davies writes:

At rock bottom, quantum mechanics provides a highly successful procedure for predicting the results of observations of microsystems, but when we ask what actually happens when an observation takes place, we get nonsense! Attempts to break out of this paradox range from the bizarre, such as the many universes interpretation of Hugh Everett, to the mystical ideas of John von Neumann and Eugene Wigner, who invoke the observer's consciousness. After half a century of argument, the quantum observation debate remains as lively as ever. The problems of the physics of the very small and the very large are formidable, but it may be that this frontier—the interface of mind and matter—will turn out to be the most challenging legacy of the New Physics.[44]

This "interface between mind and matter" is also at the core of the time paradox. If the arrow of time existed only because our human consciousness interfered with a world otherwise ruled by time-symmetrical laws, the very acquisition of knowledge would become paradoxical, since *any measure already implies an irreversible process.* If we wish to learn anything at all about a time-reversible object, we cannot avoid the irreversible processes involved in measurement, whether at the level of an apparatus or of our own sensory mechanisms. Thus, in classical physics, when we ask how we can understand "observation" in terms of fundamental time-reversible laws, we get "nonsense," as Davies puts it. In classical physics, this intrusion of irreversibility was perceived as a minor problem. The great success of classical dynamics left no doubt about its objective character. The situation is quite different in quantum theory. Here the need to include measurement in our fundamental description of nature is explicitly asserted in the very structure of the theory. It therefore seems that we have an irreducible duality: on the one hand, the time-reversible Schrödinger equation, and on the other, the collapse of the wave function.

This dualistic nature of quantum mechanics was repeatedly emphasized by the great physicist Wolfgang Pauli. In a letter to Markus Fierz in 1947, he wrote, "Something only really happens when an observation is made, and in conjunction with that . . . entropy necessarily increases. Between observations, nothing at all happens."[45] Still, the paper on which we write ages and becomes yellow, whether or not we observe it.

How can this paradox be solved? There have been many proposals put forth in addition to the extreme positions mentioned by Davies, including Niels Bohr's "Copenhagen interpretation."* Bohr concluded that the measurement apparatus has to be treated classically. It is as if we, who belong to the macroworld, need an intermediary to communicate with the microworld, just as in some religions we need a priest or shaman to communicate with the *other* world.

But this hardly solves the problem, as the Copenhagen interpretation does not lead to any prescription of what should characterize the physical systems we may use as a measurement device. Bohr avoids the basic question: What kind of dynamical processes are responsible for the collapse of the wave function? Léon Rosenfeld, Bohr's closest coworker, was quite conscious of the limitations of the Copenhagen interpretation. He considered it only a first step, the next being to give a dynamical interpretation of

---

*We highly recommend Rae, *Quantum Physics,* and A. Shimony, "Conceptual Foundations of Quantum Mechanics," in Davies' *New Physics.*

the role of the apparatus. His conviction led to a number of publications in common with our own research group, which anticipated our present approach.[46]

Other physicists have proposed identifying the measuring instrument with some "macroscopic" device. In their minds, the concept of such a device is associated with approximations. For practical reasons, we would be unable to measure the quantum properties of the apparatus. Furthermore, it has often been suggested that we should consider the apparatus as an "open" quantum system connected to the entire world.[47] Contingent perturbations and fluctuations stemming from the environment would be responsible for our ability to perform measurements. But what is meant by "environment"? Who makes the distinction between an object and its environment? This distinction is only a modified version of the von Neumann proposal, which states that through our actions and observations, it is we who produce the collapse of the wave function.

The need to eliminate the subjective element associated with the observer has been stressed by John Bell in his excellent book, *Speakable and Unspeakable in Quantum Mechanics.*[48] It is also an important consideration in the recent work of Murray Gell-Mann and James B. Hartle, who argue that the appeal to an observer becomes even more obscure in connection with cosmology.[49] Who measures the universe? This is not the place for a detailed discussion of their approach; nevertheless, a brief description of their latest findings would seem to be in order.

Gell-Mann and others introduce a coarse-grained description of the quantum mechanical histories of the universe that transforms the structure of quantum mechanics, leading from a theory of probability amplitudes to a theory of probabilities proper. As an example, let us again consider the wave function $\Psi = c_1 u_1 + c_2 u_2$ obtained by the superimposition of the wave functions u_1 and u_2. If we then take the square (for purposes of simplification, we may suppose Ψ is real) we have $\Psi^2 = c^2_1 u^2_1 + c^2_2 u^2_2 + 2c_1 c_2 u_1 u_2$. Let us now presume that we can ignore the double product called the "interference term." All the mystery of quantum theory then disappears. The probability Ψ^2 is "simply" the sum of probabilities. There is no longer any need to speak of the transition from potentiality to actuality, and we can work directly with probabilities. But how is this possible? Interference terms play a central role in many applications of quantum theory. Still, suppressing the interference term is precisely what Gell-Mann and his colleagues propose. Why then, in some situations, do we need exact, fine-grained quantum descriptions, including interference, and in others, coarse-grained ones suppressing interferences? Again, who actually does the coarse graining? Is it in any way reasonable to discuss the solution of fundamental problems in terms of *approximations?* How is this consistent with Gell-Mann's own statement, already quoted in Section II, that quantum mechanics is the framework into which all theory must fit?

Still others in the field hope to solve the quantum mechanical puzzle by reintroducing the Epicurus clinamen in a modern form. Indeed, Giancarlo Ghirardi, Emanuele Rimini, and Tullio Weber suppose that at some time, for some unknown reason, a spontaneous collapse of the wave function occurs.[50] Here the concept of chance enters the discussion, but without any deeper justification as a *deus ex machina.* Why does this new clinamen apply to some situations and not to others?

What is especially unsatisfactory about all these attempts to elucidate the conceptual foundations of quantum theory is that they make no new predictions that can actually be tested.

Our own conclusion coincides with that of many other specialists such as Abner Shimony in the United States and Bernard d'Espagnat in France.[51] According to them, radical innovations have to be made that would preserve all the achievements of quantum mechanics, but eliminate the difficulties related to the theory's dualistic structure. Note that the measurement problem is not isolated. As emphasized by Léon Rosenfeld, measurement is associated with irreversibility. But in quantum mechanics, there is no place for irreversible processes, whether or not they are involved with measurement. The difficulty of introducing irreversibility into quantum theory was already established decades ago (in the context of ergodic theory) by von Neumann, Pauli, and Fierz.[52] As in classical mechanics, they tried to solve the problem by coarse graining, but their attempts remained unsuccessful. This may be the reason that von Neumann eventually adopted a dual formulation: the Schrödinger equation on one side, and the collapse of the wave function on the other.[53] But this is hardly satisfactory as long as the collapse is not described in dynamical terms. This is precisely what our own theory achieves. The central role is again played by instability. However, deterministic chaos guided by exponentially diverging trajectories is not applicable here. In quantum mechanics, there are no trajectories. Therefore, we have to consider instability in terms of Poincaré resonances.

We can incorporate Poincaré resonances into a statistical description and derive diffusive terms that lie outside the range of quantum mechanics in terms of wave functions. The description is once again based on the level of probability ρ (also called the density matrix in quantum mechanics; see Chapter 6) and no longer on wave functions. Through Poincaré resonances, we achieve the transition from probability amplitudes to probability proper without drawing on any nondynamical assumptions.

As in classical dynamics, the basic question is, When are these diffusive terms observable? What are the limits of traditional quantum theory? The answer is similar to that for classical dynamics (see Section III). In short, it is in *persistent* interactions that the diffusive terms become dominant (see Chapter 7). As in classical mechanics, this prediction has been verified by numerical simulations. Only by going beyond a reductionist description can we give a realistic interpretation of quantum theory. There is no collapse of the wave function, as the dynamical laws are now at the level of ρ, the density matrix, and not of wave functions Ψ. Moreover, the observer no longer plays any special role. The measurement device has to present a broken time symmetry. For these systems, there is a privileged direction of time, exactly as there is a privileged direction of time in our perception of nature. It is this *common* arrow of time that is the necessary condition of our communication with the physical world; it is the basis of our communication with our fellow human beings.

Thus, instability plays a central role in both classical and quantum mechanics, and as such, obliges us to extend the scope of both disciplines. In so doing, we have to leave the field of simple integrable systems. The possibility of a unified formulation of quantum theory is particularly exciting because this problem has been so hotly debated over the past decades, but the need for an extension of classical theory is even more unexpected. We recognize that this means a break with a rational tradition that harks back to the very foundations of Western science as conceived by

Galileo and Newton. But it is no mere coincidence that the application of recent mathematical methods to unstable systems leads precisely to the extensions defined in this book. They allow us to include a description of the evolutionary characteristics of our universe based on a probabilistic description of nature. In a recent article, I. Bernard Cohen spoke of the probabilistic revolution as a revolution in applications. He wrote, "Even if the decades 1800–1930 do not show a single revolution in the domain of probability, they provide evidence of a *probabilizing revolution,* that is, of a true revolution of fantastic consequences attendant on the introduction of probability and statistics into areas that have undergone revolutionary changes as a result."[54] This "probabilizing revolution" is still going on.

V

We now come to the close of this chapter. We began with Epicurus and Lucretius, and their invention of the clinamen to permit the appearance of novelty. After twenty-five hundred years, we can at last give a precise physical meaning to this concept, which originates in instabilities identified by the modern theory of dynamical systems. If the world were formed by stable dynamical systems, it would be radically different from the one we observe around us. It would be a static, predictable world, but we would not be here to make the predictions. In our world, we discover fluctuations, bifurcations, and instabilities at all levels. Stable systems leading to certitudes correspond only to idealizations, or approximations. Curiously, this insight was anticipated by Poincaré. In discussing the laws of thermodynamics he wrote,

These laws can have only one significance, which is that there is a property common to all possibilities; but in the deterministic hypothesis there is only a single possibility, and the laws no longer have any meaning. In the indeterministic hypothesis, on the other hand, they would have meaning, even if they were taken in an absolute sense; they would appear as a limitation imposed upon freedom. But these words remind me that I am digressing and am on the point of leaving the domains of mathematics and physics.[55]

Today we are not afraid of the "indeterministic hypothesis." It is the natural outcome of the modern theory of instability and chaos. Once we have an arrow of time, we understand immediately the two main characteristics of nature: its unity and its diversity: unity, because the arrow of time is common to all parts of the universe (your future is my future; the future of the sun is the future of any other star); diversity, as in the room where I write, because there is air, a mixture of gases that has more or less reached thermal equilibrium and is in a state of molecular disorder, and there are the beautiful flowers arranged by my wife, which are objects far from equilibrium, highly organized thanks to temporal, irreversible, nonequilibrium processes. No formulation of the laws of nature that does not take into account this constructive role of time can ever be satisfactory.

Chapter 2

ONLY AN ILLUSION?

I

The results presented in this book have matured slowly. It is now more than fifty years since I published my first paper on nonequilibrium thermodynamics, in which I pointed out the constructive role of irreversibility.[1] To my knowledge, this was also the first paper that dealt with self-organization as associated with distance from equilibrium. After so many years, I often wonder why I was fascinated with the problem of time, and why it took so very long to establish its relationship with dynamics. While this is not the place to discuss the history of thermodynamics and statistical mechanics over the past half century, I do want to explain my own motivations, and indicate some of the main difficulties I encountered along the way.

I have always considered science to be a dialogue with nature. As in a real dialogue, the answers are often unexpected—and sometimes astonishing.

As an adolescent, I was enchanted with archaeology,

philosophy, and especially music. My mother used to say that I could read music before reading books. When I entered the university, I spent much more time at the piano than in lecture halls. In all the subjects I enjoyed, time played an essential role, whether in the gradual emergence of civilizations, the ethical problems associated with human freedom, or the temporal organization of sounds in music. Then came the threat of war. It seemed more appropriate to undertake a career in hard sciences, and so I started to study physics and chemistry at the Free University of Brussels.

I often questioned my teachers about the meaning of time, but their answers were conflicting. For the philosophers, this was the most difficult problem of all, closely related to ethics and the very nature of human existence. The physicists found my question somewhat naive, as the answer had already been given by Newton, and later improved upon by Einstein. As a consequence, I felt both astonished and frustrated. In science, time was considered a mere geometrical parameter. In 1796, more than one hundred years before Albert Einstein and Hermann Minkowski, Joseph–Louis Lagrange had called dynamics a "four-dimensional geometry."[2] Einstein went on to say, "Time [as associated with irreversibility] is an illusion." With my own background, these statements were impossible for me to accept. Nevertheless, the tradition of spatialized time is still very much alive today, as witnessed by the work of scientists such as Stephen W. Hawking.[3] In his *Brief History of Time,* Hawking introduces "imaginary time" to eliminate any distinction between space and time, a concept we shall examine in greater depth in Chapter 8.

I am certainly not the first to have felt that the spatialization of time is incompatible with both the evolving universe, which we observe around us, and our own human experience. This was the starting point for the French philosopher Henri Bergson, for whom "time is invention or nothing at all."[4] In Chapter 1, I mentioned one of Bergson's later articles, "The Possible and the Real," published on the occasion of his Nobel Prize in 1930, where he expressed his feeling that human existence consists of "the continual creation of unpredictable novelty," concluding that time proves that there is *indetermination* in nature.[5] The universe around us is only one of a number of possible worlds. Bergson would have been quite amazed to read Henri Poincaré's quotation at the end of Chapter 1.[6] Curiously, though, their conclusions pointed in the same direction. I also quoted Alfred North Whitehead from *Process and Reality,* for whom the ultimate goal was to reconcile permanence and change, to conceive of existence as a process. According to him, classical science, which originated in the seventeenth century, was an example of misplaced concreteness unable to express creativity as the basic property of nature, "whereby the actual world has its character of temporal passage to novelty." Whitehead's conception of the actual world was obviously incompatible with any deterministic description.[7]

I could go on by quoting Martin Heidegger and others, including Arthur Stanley Eddington, who wrote, "In any attempt to bridge the domains of experience belonging to the spiritual and physical sides of our nature, time occupies the key position."[8] But instead of building this bridge, time has remained a controversial issue from the pre-Socratics to the present day. As mentioned for classical science, the problem of time had been solved by Newton and Einstein, but for most philosophers, this solution was incomplete. In their opinion, we had to turn to metaphysics.

My personal conviction was quite different. Abandoning science appeared to be too heavy a price to pay. After all, science had led to a unique and fruitful dialogue between mankind and nature. Perhaps classical science could indeed limit time to a geometrical parameter because it was dealing only with simple problems. There was no need to extend the concept of time when we dealt with a frictionless pendulum, for instance. But once science encountered complex systems, it would have to modify its approach to time. An example that often came to mind was associated with architecture. There is not much difference between an Iranian brick from the fifth century before Christ and a neogothic brick from the nineteenth century, but the results—the palaces of Persepolis and the neogothic churches—are in striking contrast. Time would then be an "emerging" property. But what could be the roots of time? I became convinced that macroscopic irreversibility was the manifestation of the randomness of probabilistic processes on a microscopic scale. What then was the origin of this randomness?

With these preoccupations, it was only natural that I turn to thermodynamics, especially because in Brussels there was already an established school in the subject founded by Théophile De Donder (1870–1957).

II

In Chapter 1, we mentioned the classical formulation of the second law of thermodynamics attributed to Clausius. This law is based on an inequality: The entropy, S, of an isolated system increases monotonically until it reaches its maximum value at thermodynamic equilibrium. We therefore have $dS \geq 0$ for the change in entropy over the course of time. How can we extend this statement to systems that are not isolated, but which exchange energy and matter with the outside world? We must then distinguish two terms in the entropy change, dS: the first, d_eS, is the transfer of entropy across the boundaries of the system; the second, d_iS, is the entropy produced within the system. As a result, we have $dS = d_eS + d_iS$. We can now express the second law by stating that whatever the boundary conditions, the entropy production d_iS is positive, that is, $d_iS \geq 0$. *Irreversible processes are creating entropy.* De Donder went even farther: He expressed the production of entropy per unit time $P = \frac{d_iS}{dt}$ in terms of the rates of various irreversible processes (chemical reaction rates, diffusion, etc.) and thermodynamic forces. In fact, he considered only chemical reactions, but further generalization was easy.[9]

De Donder himself did not go very far along this road. He was concerned mainly with equilibrium and the neighborhood of equilibrium. Limited as it was, his work represented an important step in the formulation of non-equilibrium thermodynamics, even if it seemed to lead nowhere for a considerable length of time. I still remember the hostility with which De Donder's work was met. For the vast majority of scientists, thermodynamics had to be limited *strictly* to equilibrium.

That was the opinion of J. Willard Gibbs, as well as of Gilbert N. Lewis, the most renowned thermodynamicist of his day. For them, irreversibility associated with unidirectional time was anathema. Lewis went so far as to write, "We shall see that nearly everywhere the physicist has purged from his science the use of one-way time . . . alien to the ideals of physics."[10]

I myself experienced this type of hostility in 1946, when I organized the first Conference on Statistical

Mechanics and Thermodynamics under the auspices of the International Union for Pure and Applied Physics (IUPAP). These meetings have since been held on a regular basis and continue to attract large crowds, but at that time we were a small group of approximately thirty to forty people. After I had presented my own lecture on irreversible thermodynamics, the greatest expert in the field of thermodynamics made the following comment: "I am astonished that this young man is so interested in nonequilibrium physics. Irreversible processes are transient. Why not wait and study equilibrium as everyone else does?" I was so amazed at this response that I did not have the presence of mind to answer: "But we are all transient. Is it not natural to be interested in our common human condition?"

Throughout my entire life I have encountered hostility to the concept of unidirectional time. It is still the prevailing view that thermodynamics as a discipline should remain limited to equilibrium. In Chapter 1, I mentioned the attempts to banalize the second law that are so much a part of the credo of a number of famous physicists. I continue to be astonished by this attitude. Everywhere around us we see the emergence of structures that bear witness to the "creativity of nature," to use Whitehead's term. I have always felt that this creativity had to be connected in some way to the distance from equilibrium, and was thus the result of irreversible processes.

Compare, for example, a crystal and a town. A crystal is an equilibrium structure that can be maintained in a vacuum, but if we isolated the town, it would die because its structure depends on its function. Function and structure are inseparable in that the latter expresses the interactions of the town with its environment.

In Erwin Schrödinger's beautiful book *What Is Life?* he discusses the metabolism of a living body in terms of entropy production and entropy flow. If an organism is in a steady state, its entropy remains constant over time, and therefore $dS = 0$. As a result, the entropy production $d_i S$ is compensated by the entropy flow $d_i S + d_e S = 0$, or $d_e S = d_i S < 0$. Life, concludes Schrödinger, feeds on a "negative entropy flow."[11] The more important point, however, is that life is associated with entropy production and therefore with irreversible processes.

But how can structure, as in living systems or towns, emerge in nonequilibrium conditions? Here again, as in dynamics, the problem of *stability* plays an essential role. At thermodynamic equilibrium, entropy has a maximum value when the system is isolated. For a system maintained at temperature T, we have a similar situation. We then introduce "free energy," $F = E - TS$, a linear combination of energy E and entropy S. As shown in all texts on thermodynamics, free energy, F, is at its minimum at equilibrium (see Figure 2.1). Consequently, perturbations or fluctuations have no effect because they are followed by a return to equilibrium. The situation is not unlike that of the stable pendulum considered in Chapter 1, Section III.

What happens in a steady state corresponding to nonequilibrium? We saw such an example in a discussion of thermal diffusion in Chapter 1, Section II. Is a nonequilibrium steady state truly stable? In near-equilibrium situations (known as "linear" nonequilibrium thermodynamics), the answer is yes. As shown in 1945, the steady state corresponds to a minimum of entropy production per unit time $P = \frac{d_i S}{dt}$.[12] At equilibrium $P = 0$, entropy production vanishes, while in the linear regime around equilibrium, P is minimum (see Figure 2.2).[13]

Figure 2.1

Minimum of *F*

Free energy is minimum at equilibrium ($\lambda = \lambda_m$).

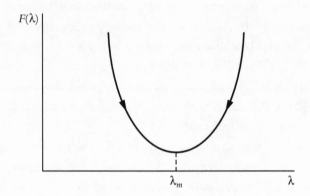

Again fluctuations die out. But there already appears a remarkable new characteristic: A nonequilibrium system may evolve spontaneously to a state of *increased complexity*. The ordering we observe is the outcome of irreversible processes, and could not be achieved at equilibrium. This is clear in the example of thermal diffusion mentioned in Chapter 1, where the temperature gradient leads to a partial separation of the compounds. Many other cases have since been studied in which complexity has consistently been associated with irreversibility. These results became the guidelines for our future research.

But can we extrapolate the results of far-from-equilibrium situations from those at near-equilibrium? My colleague Paul Glansdorff and I investigated this problem for many years,[14] and arrived at a surprising conclusion: Contrary to what happens at equilibrium, or near equilibrium, systems far from equilibrium do not conform to any minimum principle that is valid for functions of free energy or entropy production. As a consequence, there is no guarantee that fluctuations are damped. We can only achieve

Figure 2.2

Minimum of *P*

Entropy production $P = d_i S/dt$ is minimum in a steady state ($\lambda = \lambda_{st}$).

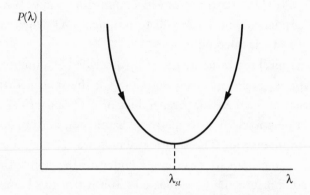

a formulation of *sufficient conditions* for stability, which we call the "general evolution criterion." This requires specifying the mechanism of irreversible processes. Near-equilibrium laws of nature are *universal*, but when they are far from equilibrium, they become mechanism dependent. We therefore begin to perceive the origin of the variety in nature we observe around us. Matter acquires new properties when far from equilibrium in that fluctuations and instabilities are now the norm. Matter becomes more "active." Although there is at present an enormous literature surrounding this subject,[15] for the moment we shall consider only a simple example. Suppose that we have a chemical reaction $\{A\} \rightleftharpoons \{X\} \rightleftharpoons \{F\}$ in which $\{A\}$ is a set of initial products, $\{X\}$ a set of intermediate ones, and $\{F\}$ a set of final ones. At equilibrium, we have a detailed balance where there are as many transitions from $\{A\}$ to $\{X\}$ as from $\{X\}$ to $\{A\}$, with the same applying to $\{X\}$ and $\{F\}$. The ratio of initial to final products $\{A\}/\{F\}$ takes on a well-defined value corresponding to maximum entropy if the system is isolated. Now consider an open system, such as a chemical reactor. By controlling the flow of matter, we may fix the values of both the initial and final products $\{A\}$ and $\{F\}$. We progressively increase the ratio $\{A\}/\{F\}$, starting from its equilibrium value. What will happen to the intermediate products $\{X\}$ when we move away from equilibrium?

Chemical reactions are generally described by nonlinear equations. There are many solutions for the intermediate concentrations $\{X\}$ for given values of $\{A\}$ and $\{F\}$, but only one corresponds to thermodynamic equilibrium and maximum entropy. This solution, which we call the "thermodynamic branch," may be extended to the domain of nonequilibrium. The unexpected result is that this branch generally becomes *unstable* at some critical distance from equilibrium (see Figure 2.3). The point where this occurs is known as the bifurcation point.

Beyond the bifurcation point, a set of new phenomena arises; we may have oscillating chemical reactions, non-equilibrium spatial structures, or chemical waves. We have given the name *dissipative structures* to these spatiotemporal organizations. Thermodynamics leads us to the formulation of two conditions for the occurrence of dissipative structures in chemistry: (1) far-from-equilibrium situations defined by a critical distance; and (2) catalytic steps, such as the production of the intermediate compound Y from compound X *together* with the production of X from Y.

It is interesting to note that these conditions are satisfied in all living systems: Nucleotides code for proteins, which in turn code for nucleotides.

We were extremely fortunate in that soon after we had predicted these various possibilities, the experimental results of the Belousov-Zhabotinski reaction—a spectacular example of chemical oscillations—became widely known.[16] I remember our amazement when we saw the reacting so-

Figure 2.3

Thermodynamic Branch

The two steady-state solutions *th* and *d* are functions of the ratio *A/F*. At the bifurcation point, the thermodynamic branch *th* becomes unstable, and another branch *d* becomes stable.

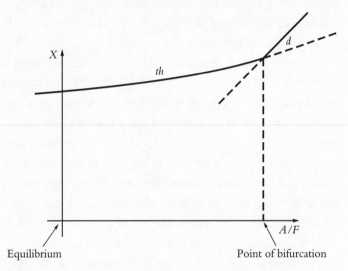

lution become blue, and then red, and then blue again. Today, many other oscillatory reactions are known,[17] but the Belousov-Zhabotinski reaction remains historic because it proved that matter far from equilibrium acquires new properties. Billions of molecules become simultaneously blue, and then red. This entails the appearance of long-range correlations in far-from-equilibrium conditions that are absent in a state of equilibrium. Again, we can say that matter at equilibrium is "blind," but far from equilibrium it begins to "see." We have observed that at near equilibrium, dissipation associated with entropy production is at a minimum. Far from equilibrium, it is just the opposite. New processes set in and increase the production of entropy.

There has been steady progress in far-from-equilibrium chemistry. In recent years, nonequilibrium spatial structures have been observed.[18] These were first predicted by Alan Mathison Turing in the context of morphogenesis.[19] When we push the system farther into nonequilibrium, new bifurcations typical of chaotic behavior may arise. Neighboring trajectories diverge exponentially as in deterministic chaos akin to the dynamical systems we considered in Chapter 1, Section III.

In short, distance from equilibrium becomes an essential parameter in describing nature much like temperature in equilibrium thermodynamics. When we lower the temperature, we observe a succession of phase transitions through various states of matter. But in nonequilibrium physics, the variety of behaviors is much greater. We have considered chemistry for the purposes of this discussion, but similar processes associated with nonequilibrium dissipative structures have been studied in many other fields, including hydrodynamics, optics, and liquid crystals.

Let us now look more closely at the critical effect of fluctuations. As we have seen, near-equilibrium fluctuations are harmless, but far from equilibrium, they play a central role. Not only do we need irreversibility, but we also have to abandon the deterministic description associated with dynamics. The system "chooses" one of the possible branches available when far from equilibrium. But nothing in the macroscopic equations justifies the preference for any one solution. This introduces an irreducible probabilistic element. One of the simplest bifurcations is the so-called "pitchfork bifurcation" represented in Figure 2.4, where $\lambda = 0$ corresponds to equilibrium.

The thermodynamic branch is stable from $\lambda = 0$ to $\lambda = \lambda_c$. Beyond λ_c, it becomes unstable, and a symmetrical pair of new stable solutions emerges. It is the fluctuations that decide which branch will be selected. If we were to suppress fluctuations, the system would maintain itself in

Figure 2.4

Pitchfork Bifurcation

Concentration X is a function of the parameter λ, which measures the distance from equilibrium. At the bifurcation point, the thermodynamic branch becomes unstable, and the two new solutions b_1 and b_2 emerge.

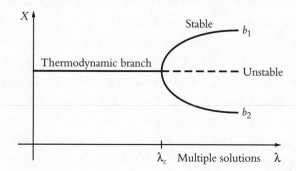

an unstable state. Attempts have been made to decrease the fluctuations so that we can subject the unstable region to experiment; nevertheless, sooner or later, fluctuations of internal or external origin take over and bring the system to one of the branches b_1 or b_2.

Bifurcations are a source of symmetry breaking. In fact, the solutions of the equation beyond λ_c generally have a lower symmetry than the thermodynamic branch.[20] Bifurcations are the manifestation of an intrinsic differentiation between parts of the system itself and the system and its environment. Once a dissipative structure is formed, the homogeneity of time (as in oscillatory chemical reactions) or space (as in nonequilibrium Turing structures), or both, is broken.

In general, we have a succession of bifurcations as represented schematically in Figure 2.5. The temporal description of such systems involves both deterministic processes (between bifurcations) and probabilistic processes (in the choice of the branches). There is also a historical dimension involved. If we observe that the system is in state d_2, that means that it has gone through the states b_1 and c_1 (see Figure 2.5).

Once we have dissipative structures, we can speak of self-organization. Even if we know the initial values and boundary constraints, there are still many states available to the system among which it "chooses" as a result of fluctuations. Such conclusions are of interest beyond the realms of physics and chemistry. Indeed, bifurcations can be considered the source of diversification and innovation.[21] These concepts are now applied to a wide group of problems in biology, sociology, and economics at interdisciplinary centers throughout the world. In Western Europe alone, there have been more than fifty centers for nonlinear processes founded over the past ten years.

Freud wrote that the history of science is the history of alienation. Copernicus showed that the earth is not at the center of the planetary system, Darwin that we are one species of animal among many others, and Freud that our

Figure 2.5
Successive Bifurcations with Increasing Distance from Equilibrium

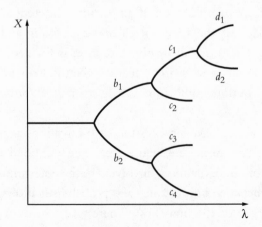

rational activity is only part of the unconscious. We can now invert this perspective: We see that human creativity and innovation can be understood as the amplification of laws of nature already present in physics or chemistry.

III

The results presented thus far show that the attempts to trivialize thermodynamics mentioned in Chapter 1 are necessarily doomed to failure. The arrow of time plays an essential role in the formation of structures in both the physical sciences and biology. But we are only at the beginning of our quest. There is still a gap between the most complex structures we can produce in nonequilibrium situations in chemistry and the complexity we find in biology. This is not only a problem for pure science. In a recent report to the European Communities, Christof Karl Biebracher, Grégoire Nicolis, and Peter Schuster wrote,

The maintenance of organization in nature is not—and cannot be—achieved by central management; order can only be maintained by self-organization. Self-organizing systems allow adaptation to the prevailing environment, i.e., they react to changes in the environment with a thermodynamic response which makes the systems extraordinarily flexible and robust against perturbations from outside conditions. We want to point out the superiority of self-organizing systems over conventional human technology which carefully avoids complexity and hierarchically manages nearly all technical processes. For instance, in synthetic chemistry, different reaction steps are usually carefully separated from each other and contributions from the diffusion of the reactants are avoided by stirring reactors. An entirely new technology will have to be developed to tap the high guidance and regulation potential of self-organizing systems for technical processes. The superiority of self-organizing systems is illustrated by biological systems where complex products can be formed with unsurpassed accuracy, efficiency and speed!"[22]

The results of nonequilibrium thermodynamics are close to the views expressed by Bergson and Whitehead. Nature is indeed related to the creation of unpredictable novelty, where the possible is richer than the real. Our universe has followed a path involving a succession of bifurcations. While other universes may have followed other paths, we are fortunate that ours has led to life, culture, and the arts.

The dream of my youth was to contribute to the unification of science and philosophy by resolving the enigma of time.* Nonequilibrium physics shows that this is entirely possible. The results described in this chapter gave me the impetus to explore the concept of time on the microscopic level. I have emphasized the role of fluctuations—but what is their origin? How can we reconcile their behavior with the deterministic description based upon the traditional formulation of the laws of nature? Were we to do so, we would lose the distinction between near- and far-from-equilibrium processes. Moreover, we would be calling into question such unique and marvelous constructions of the human mind as classical and quantum mechanics.

I must confess that these thoughts led to many sleepless nights. Without the support of my colleagues and students, I would most certainly have given up.

---

*I expressed this dream in three short essays written for a student journal as early as 1937!

Chapter 3

FROM PROBABILITY TO IRREVERSIBILITY

I

As we saw in Chapter 2, irreversible processes describe fundamental features of nature leading to nonequilibrium dissipative structures. Such processes would not be possible in a world ruled by the time-reversible laws of classical and quantum mechanics. Dissipative structures require an arrow of time. Furthermore, there is no hope of explaining the appearance of such structures through approximations that would be introduced by these laws.

I have always been convinced that an understanding of the dynamical origin of dissipative structures, and more generally of complexity, is one of the most fascinating conceptual problems of contemporary science. As already stated in Chapter 1, for unstable systems we have to formulate the laws of dynamics at the statistical level. This changes our description of nature in a radical way. In such a formulation, the basic objects of physics are no longer trajectories or wave functions; they are *probabilities*. We have thus come to the end of the "probabilistic revolution" that could already be found in areas other than physics by the eighteenth century. However, when faced with the implications of this radical conclusion, I hesitated for some time, reaching for less extreme solutions. In *From Being to Becoming,* I wrote, "In quantum mechanics, there are observations whose numerical value cannot be determined simultaneously, i.e., coordinates and momentum. (This is the essence of Heisenberg's uncertainty relations and Bohr's complementarity principle.) Here we also have a complementarity—one between dynamical and thermodynamical descriptions."[1] This would have been a much less extreme approach to the conceptual problem associated with irreversibility.

In retrospect, I regret this statement in my earlier book. If there is more than a single description, who would choose the right one? The existence of the arrow of time is not a matter of convenience. It is a fact imposed by observation. However, it is only in recent years that the results we obtained by studying the dynamics of unstable systems forced us to reformulate dynamics at the statistical level, and to conclude that this formulation leads to an extension of classical and quantum mechanics. In this chapter, I describe some of the steps involved.

For approximately one hundred years, we have known that even simple probabilistic processes are time oriented. In Chapter 1, we mentioned the "random walk." Another example is the "urn model" proposed by Paul and Tatiana Ehrenfest (see Figure 3.1).[2]

Consider N objects (such as balls) distributed between

Figure 3.1

The Ehrenfest Urn Model

N balls are distributed between two urns, A and B. At time *n*, there are *k* balls in A and *N* − *k* balls in B. At regular time intervals, a ball is removed at random from one urn and placed into the other.

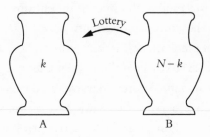

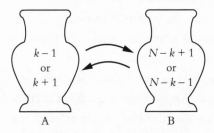

two urns *A* and *B*. At regular time intervals (for example, every second) a ball is chosen at random and moved from one urn to the other. Suppose that at time *n*, there are *k* balls in *A*, and therefore *N* − *k* in *B*. At time *n* + *1* there can be either *k* − *1* or *k* + *1* balls in *A*. These are well-defined *transition probabilities*. But let us go on with the game. We expect that as a result of the exchange of balls, we shall reach a situation where there will be approximately $\frac{N}{2}$ balls in each urn. However, fluctuations will continue. We might even end up in the situation at time *n* where there are again *k* balls in urn *A*. It is at the level of *probability distribution* that we see an irreversible approach to equilibrium. Whatever the starting point, it can be shown that the probability $p_n(k)$ of finding k balls in one urn after n moves as $n \to \infty$ tends to the binomial distribution $\frac{N!}{k!(N-k)!}$. This expression has a maximum value of $k = \frac{N}{2}$, but also takes into account fluctuations in distribution. In the Boltzmann model, the maximum entropy corresponds precisely to the binomial distribution.

The Ehrenfest model is an example of a "Markov process" (or "Markov chain"), named after the great Russian mathematician, Andrei Markov, who was the first to describe such processes. Once we have a probabilistic description, it is often possible to derive irreversibility. But how do we relate these probabilistic processes to dynamics? That is the fundamental problem.

We have seen that a basic step in this direction was taken by the fathers of statistical physics, or the physics of populations. Maxwell, Boltzmann, Gibbs, and Einstein all emphasized the role of ensembles described by a probability distribution ρ. An important question then is, What is the form of this distribution function once equilibrium is reached? Let $q_1, \ldots, q_s$ be the coordinates and $p_1, \ldots, p_s$ the momenta of the particles forming this system. In Chapter 1, the phase space was defined by the coordinates and momenta. We also introduced the probability distributions $\rho(q, p, t)$ (see Chapter 1, Section III). We shall now use the single letter q for all coordinates and p for all momenta. Equilibrium is reached when ρ becomes time independent. In every textbook, it is shown that this occurs when ρ depends only on the total energy. As mentioned in Chapter 1, Section III, the total energy is the sum of the kinetic energy (due to the motion of the particles) and the potential energy (due to interactions). When expressed in terms of q and p, this energy, which is called the *Hamiltonian* $H(p, q)$, remains constant over time. This is the principle of conservation of energy, the first principle of thermodynamics. It is therefore natural that at equilibrium, ρ is a function of the *Hamiltonian H.*

An important exception case is that of ensembles in which all systems have the same energy E. The distribution function then vanishes throughout the phase space, save on the surface $H (p, q) = E,$ where the distribution function is constant. This is called the "microcanonical ensemble." Gibbs showed that such ensembles do indeed satisfy the laws of equilibrium thermodynamics. He also considered other ensembles such as the "canonical ensemble," in which all systems interact with a reservoir at temperature T. This leads to a distribution function that depends exponentially on the Hamiltonian, ρ now being proportional to exp $(- \frac{H}{kT})$, where T is the temperature of the reservoir and k the Boltzmann constant, which makes the exponent dimensionless.

Once the equilibrium distribution is given, we can calculate all thermodynamic equilibrium properties such as pressure, specific heat, etc. We can even go beyond macroscopic thermodynamics because we are able to include fluctuations. It is generally accepted that in the vast field of equilibrium statistical thermodynamics, there are no conceptual difficulties left, only computational ones, which can be solved largely through numerical simulations. The application of ensemble theory to equilibrium situations has undoubtedly been quite successful. Note that the dynamical interpretation of equilibrium thermodynamics by Gibbs is in terms of *ensembles,* and not in terms of trajectories. It is this approach that we have to extend in order to include irreversibility.

This is quite natural, as there is no time ordering at the level of trajectories (or wave functions) because future and past play the same role in accordance with classical and quantum physics. However, what happens at the level of statistical description, in terms of distribution functions? Let us look at a glass of water. In this glass, there is a huge number of molecules, a quantity on the order of 10^{23}. From the dynamical point of view, this is a nonintegrable Poincaré system, as defined in Chapter 1, since there are interactions between the molecules that we cannot eliminate. We may visualize these interactions as leading to collisions between the molecules (the term "collision" will be defined more precisely in Chapter 5), and describe the water containing them in terms of the statistical ensemble ρ. Is the water aging? Certainly not, if we consider the individual water molecules, which are stable over geological time. Still, there is a natural time order in this system from the point of view of the statistical description. Aging is a property of populations, exactly as it is in the Darwinian theory of biological evolution. It is the statistical distribution that approaches the equilibrium distribution, such as the canonical distribution defined above. To describe this approach to equilibrium, we need the idea of *correlation*.

Consider a probability distribution $\rho(x_1, x_2)$, depending on two variables x_1, x_2. If x_1 and x_2 are independent, we have the factorization $\rho(x_1, x_2) = \rho_1(x_1)\rho_2(x_2)$. The probability $\rho(x_1, x_2)$ is then the product of two probabilities. In contrast, if $\rho(x_1, x_2)$ cannot be factorized, x_1 and x_2 are *correlated*. Now let us return to the molecules in the glass of water. The collisions between these molecules have two effects: They make the velocity distribution more symmetrical, and they produce correlations (see Figure 3.2). But two correlated particles will eventually collide with a third one. Binary correlations are then transformed into ternary ones, and so on (see Figure 3.3).

We now have a flow of correlations that are ordered in

Figure 3.2

Collisions and Correlations

The collision of two particles creates a correlation between them (represented by a wavy line).

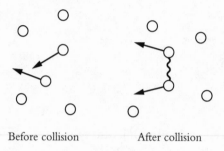

Before collision   After collision

time. A valuable and provocative analogy to this flow would be human communication. When two people meet, they converse, and consequently modify their thinking to some extent. These modifications are brought to subsequent meetings, and modified further. The word for this phenomenon is *dissemination*. There is a flow of communication in society, just as there is a flow of correlations in matter. Of course, we may also conceive of inverse processes that make the velocity distribution less symmetrical by destroying correlations (see Figure 3.4).

We therefore need an element that will validate the processes that make the velocity distribution more symmetrical over the course of time. As we shall see, this is precisely the role of Poincaré resonances. We now begin to get a glimpse of a statistical description that includes irreversibility. This description will be a *dynamics of correlations* leading to the equilibrium distribution.

The existence of a flow of correlations ordered in time, as represented in Figure 3.3, has been verified by computer simulations.[3] We can also reproduce processes such as those

Figure 3.3
Flow of Correlations

Successive collisions lead to binary, ternary, ... correlations.

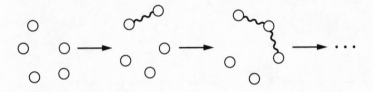

represented in Figure 3.4 through time inversion, where we invert the velocity of the particles. But we can achieve this inverted flow of correlations only for brief periods of time and for a limited number of particles, after which we again have a directed flow of correlations involving an ever-increasing number of particles leading the system to equilibrium.

Figure 3.4
Destruction of Correlations

In (a) the particles (represented by black points) interact with the obstacle (represented by the circle). Initially all particles have the same velocity. The collision varies the velocities and creates correlations between the particles and the obstacle. In (b) we represent the opposite process. We consider the effect of a velocity inversion; as a result of the inverted collision, correlations with the obstacle are destroyed, and the initial velocity is recovered.

(a)   (b)

These results, which give meaning to irreversibility at the statistical level, were obtained nearly thirty years ago.[4] At that time, however, certain basic questions still remained unanswered: How can irreversibility appear at the statistical level of description, and not when we describe dynamics in terms of trajectories? Is this due to our approximations? Moreover, is the succession of correlations that we observe, for example in computer experiments, perhaps the result of the limitations of computer time? Obviously, a shorter program is required to prepare uncorrelated particles that produce correlations through collisions than to prepare ensembles that could lead to inverse processes in which correlations are destroyed.

But why start at all with probability distributions? Such distributions describe the behavior of bundles of trajectories, or ensembles. Do we use ensembles because of our "ignorance," or is there, as argued in Chapter 1, a deeper reason involved? For unstable systems, ensembles indeed display new properties as compared with individual trajectories. This is what we shall now demonstrate with several simple examples.

II

In this section, we shall be concerned with deterministic chaos, as well as an especially simple type of chaos, both corresponding to *chaotic maps*. Contrary to what occurs in ordinary dynamics, time in maps acts only at discrete intervals, as is the case in the Ehrenfest urn model we studied in Section I. Maps therefore represent a simplified form of dynamics that makes it easy for us to compare the individual level of description (the trajectories) with the statistical description. We shall consider two maps; the first charts simple periodic behavior, the second deterministic chaos.

In the first instance, we shall consider the "equations of motion" $x_{n+1} = x_n + \frac{1}{2}$, modulo 1, which means that we are dealing only with numbers between 0 and 1. After two shifts, we are back to the initial point (i.e., $x_o = \frac{1}{4}$, $x_1 = \frac{3}{4}$, $x_2 = \frac{3}{4} + \frac{2}{4} = \frac{5}{4} = \frac{1}{4}$). This situation is represented in Figure 3.5.

Instead of considering individual points located by trajectories, it is worth examining ensembles described by the probability distribution $\rho(x)$. A trajectory corresponds to a specific set of ensembles where the coordinate x takes on a well-defined value x_n, and the distribution function ρ is then reduced to a single point. As mentioned in Chapter 1, Section III, this can be written as $\rho_n(x) = \delta(x - x_n)$. (Delta is a symbol for a function that vanishes for all values of x

Figure 3.5

Periodic Map

There is a simple geometrical construction that moves from the initial point P_0 to the next point P_1 according to the map $x_{n+1} \rightarrow x_n + 1/2$. We go from P_0 to P', then to P'' on the bisector, and from there to P_1. Obviously, if we start with P_1, we come back to P_0.

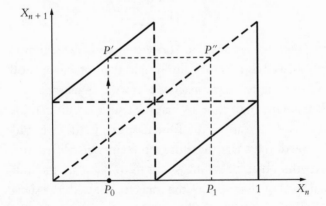

except $x = x_n$.) By using distribution function ρ, the mapping can be expressed as a relation between $\rho_{n+1}(x)$ and $\rho_n(x)$. We may then write $\rho_{n+1}(x) = U \rho_n(x)$. Formally, ρ_{n+1} is obtained through the operator U, known as the Perron-Frobenius operator, acting on $\rho_n(x)$.[5] At this point, although its explicit form is not important to us, it is interesting to note that no new element (in addition to the equation of motion) enters into the construction of U. Obviously, the ensemble description must allow the trajectory description as a special case; we therefore have $\delta(x - x_{n+1}) = U\delta(x - x_n)$. This is simply a way of rewriting the equation of motion, as x_n becomes x_{n+1} after one shift. The main question is, however, *Is this the only solution, or are there new solutions for the evolution of ensembles, as described by the Perron-Frobenius operator, which cannot be expressed in terms of trajectories?* In our example of a periodic map, the answer is no. There is not any difference between the behavior of individual trajectories and ensembles for stable systems. It is this equivalence between the individual point of view (corresponding to trajectories or wave functions) and the statistical point of view (corresponding to ensembles) that is broken for unstable dynamical systems.

The simplest example of a chaotic map is the *Bernoulli map*. Here we double the value of a number between 0 and 1 every second. The equation of motion is now $x_{n+1} = 2x_n$ (modulo 1). This map is represented in Figure 3.6. The equation of motion is again deterministic in that once we know x_n, the number x_{n+1} is determined. Here we have an example of deterministic chaos, so called because if we follow a trajectory through numerical simulations, we see that it becomes erratic. As the coordinate x is multiplied by two at each step, the distance between two trajectories will

Figure 3.6

Bernoulli Map

In this example of deterministic chaos, we start from point P_0 and go to point P_1, as the value of x doubles (modulo 1).

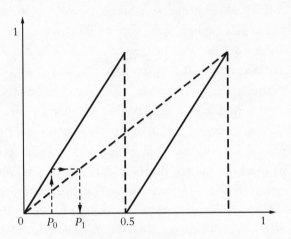

be $(2^n) = \exp(n \log 2)$, again modulo 1. In terms of continuous time t, this can be written as $\exp(t\lambda)$, with $\lambda = \log 2$, where λ is called the Lyapunov exponent. This shows that trajectories diverge exponentially, and it is this divergence that is the signature of deterministic chaos. If we wait long enough, any arbitrarily selected point between 0 and 1 will eventually be approached by the trajectory (see Figure 3.7). Here we have a dynamical process leading to randomness. In the past, this apparent flow in the deterministic universe was repeatedly investigated by great mathematicians such as Leopold Kronecker (1884) and Hermann Weyl (1916). According to Jan von Plato, similar results had been obtained as early as medieval times, so this is certainly not a new problem.[6] What is new, however, is the statistical formulation of the Bernoulli map, which links randomness to operator theory.

Figure 3.7

Numerical Simulations of Trajectories for the
Bernoulli Map

The initial conditions are slightly different for each simulation.
This difference is amplified as time goes on. (These numerical
simulations are the work of Dean Driebe).

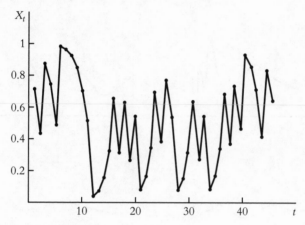

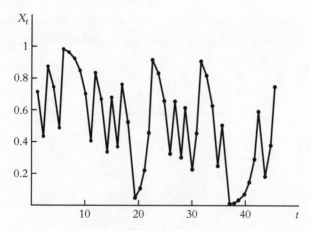

Figure 3.8

Simulation of ρn(x) for the Bernoulli Map

Numerical simulation of the evolution of the probability distribution. In contrast with the trajectory description, the probabilities rapidly reach the asymptotic uniform distribution. (These numerical simulations are the work of Dean Driebe.)

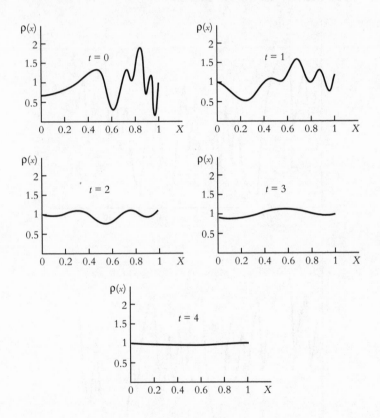

We now turn to the statistical description in terms of the Perron-Frobenius operator. In Figure 3.8, we see the effect of the operator U on the distribution function. The difference from the trajectory description is striking because the distribution function $\rho_n(x)$ leads rapidly to a constant. We may therefore conclude that there must be a basic difference between the description in terms of trajectories on the one hand and in terms of ensembles on the other. In short, instability at the level of trajectories leads to stability at the level of statistical descriptions.

How is this possible? The Perron-Frobenius operator still admits a trajectory description $\delta(x - x_{n+1}) = U\delta(x - x_n)$, but the unexpected feature is that it also allows new solutions that are applicable *only* to statistical ensembles, and not to individual trajectories. The equivalence between the individual point of view and the statistical description is broken.

This remarkable fact leads to a new chapter in mathematics and theoretical physics.[7] Although the problem of chaos cannot be solved at the level of individual trajectories, it can be solved at the level of ensembles. We can now speak of the *laws of chaos*.[8] As we shall see in Chapter 4, we may even predict the speed at which the distribution ρ approaches equilibrium (which for the Bernoulli map is a constant), and establish the relationship between this speed and the Lyapunov exponent.

How can we understand the difference between individual description and statistical description? We shall analyze this situation in more detail in Chapter 4, where we shall see that these new solutions require smoothness in the distribution functions. This is the reason that such solutions are not applicable to individual trajectories. A trajectory represented by $\delta(x - x_n)$ is not a smooth function; it is different from zero only for $x = x_n$, and vanishes if x differs at all from x_n.

The description in terms of distribution functions is therefore richer than that derived from individual trajectories. This agrees with the conclusions we arrived at in Chapter 1, Section III. Trajectories are merely special solutions of the Perron-Frobenius equation for unstable maps. This also applies to systems with Poincaré resonances (see Chapters 5 and 6). The time-oriented flow of correlations is an essential element in the new solutions for the probability distribution, while no time-oriented processes exist at the level of individual trajectories.

This break in the equivalence between the individual and the statistical description is the fundamental inspiration of our approach. In the next chapter, we shall discuss in greater detail the new solutions that arise in chaotic maps at the statistical level.

The situation which we now find ourselves in is reminiscent of the one we encountered in thermodynamics (Chapter 2). The very success of equilibrium thermodynamics has retarded the discovery of new properties of matter in nonequilibrium situations where dissipative structures and self-organization appear. In parallel, the success of classical trajectory theory and quantum mechanics has hampered the extension of dynamics to the statistical level in which irreversibility can be incorporated into the basic description of nature.

Chapter 4

THE LAWS OF CHAOS

I

In the preceding chapter, we formulated the principal factor that makes it possible for us to extend classical and quantum mechanics for unstable dynamical systems: the breaking of the equivalence between the individual description (in terms of trajectories) and the statistical description (in terms of ensembles). We now wish to analyze this inequivalence more closely for simple chaotic maps and illustrate how this observation relates to recent developments in mathematics.[1] Let us first return to the Bernoulli map, which we have already introduced as an example of deterministic chaos.

We see from the equation of motion $x_{n+1} = 2x_n \pmod 1$ that we may calculate x_n for arbitrary n once we know the initial condition x_0. However, an essential element of randomness still appears to be present. An arbitrary number x between 0 and 1 can be represented in a binary digital system: $x = \frac{u_0}{2} + \frac{u_{-1}}{4} + \frac{u_{-2}}{8} \ldots$, where $u_i = 0$ or 1 (we are using the negative indices u_{-1}, u_{-2} to introduce the baker transformation, which we shall study in Section III. Each number x_n is thus represented by a series of digits. We can easily verify that the Bernoulli map leads to the shift $u'_n = u_{n-1}$ (for instance, $u'_{-2} = u_{-3}$) as it moves the numbers u_i to the left. Because the value of each digit in the series u_{-1}, u_{-2}, . . . is independent of the others, the result of each successive shift is as random as flipping a coin. This system is called a "Bernoulli shift," in memory of the pioneering work in games of chance done by the great eighteenth-century mathematician, Jakob Bernoulli. Here we can also observe a sensitivity to initial conditions: Two numbers differing only slightly (for example, by u_{-40}, which means less than 2^{-39}) will differ by $\frac{1}{2}$ after 40 steps. As we have already explained, this sensitivity corresponds to a positive Lyapunov exponent whose value is log 2 as x doubles at each step (see Chapter 3, Section II).

From the outset, the Bernoulli map introduces an arrow of time that can only point in one direction. If, instead of $x_{n+1} = 2x_n$ (mod 1), we consider the map $x_{n+1} = \frac{1}{2}x_n$, we find a single-point attractor at $x = 0$. The time symmetry is broken at the level of the equation of motion, which is thus not invertible. This is in contrast to the dynamical systems described by Newton, whose equations of motion are invariant with respect to time inversion.

The most important point to keep in mind at this juncture is that trajectories are inadequate. They are incapable of describing the time evolution of chaotic systems even if they are governed by deterministic equations of motion. As Pierre-Maurice Duhem stated as early as 1906, the notion of trajectory is an adequate mode of representation only if the trajectory remains more or less the same when we slightly modify the initial conditions.[2] The description of chaotic systems in terms of trajectories lacks precisely this robustness. This is the very meaning of sensitivity to initial conditions: Two trajectories taking off from points as close together as we can imagine will diverge exponentially over the course of time.

On the contrary, there is no difficulty in describing chaotic systems at the statistical level. It is therefore at this level that we have to formulate the laws of chaos. In Chapter 3, we introduced the Perron-Frobenius operator U, which transforms the probability distribution $\rho_n(x)$ into $\rho_{n+1}(x)$, leading us to conclude that there exist new solutions that are not applicable to individual trajectories. It is these novel solutions that we wish to identify in this chapter. The study of the Perron-Frobenius operator, which is a rapidly growing field, is of special interest here because chaotic maps are perhaps the simplest systems that display irreversible processes.

Boltzmann applies his ideas to gases containing an immense number of particles (on the order of 10^{23}). Here, on the other hand, we are dealing with only a few independent variables (one for the Bernoulli map and two for the baker map, which we shall consider shortly). Once again, we shall have to reject the contention that irreversibility exists only because our measurements are limited to approximations. But first let us identify the new class of solutions associated with the statistical description.

II

How do we solve a dynamical problem at the statistical level? First we need to determine the distribution function $\rho(x)$ so that we can observe the recurrence relation $\rho_{n+1}(x)$

$= U \rho_n(x)$. The distribution function $\rho_{n+1}(x)$ after $(n + 1)$ maps is obtained by the action of the operator U on $\rho_n(x)$, which is the distribution function after n maps. We shall meet the same type of problem in classical and quantum mechanics. For reasons that we shall explain in Chapter 6, operator formalism was first introduced in quantum theory, and then extended to other fields of physics, most notably statistical mechanics.

An operator is simply a prescription for how to act on a given function; as such, it may involve multiplication, differentiation, or any other mathematical operation. In order to define the operator, we must also specify its domain. On what types of functions does the operator act? Are they continuous or bounded? Do they have other characteristics as well? These properties define the function space.

In general, an operator U acting on a function $f(x)$ transforms it into a different function. (For instance, if U is a derivative operator $\frac{d}{dx}$, then $Ux^2 = 2x$). However, there are special functions, known as the *eigenfunctions* of the operator, which remain invariant when we apply U; they are multiplied only by a number known as the *eigenvalue*. In the above example, e^{kx} is an eigenfunction to which the eigenvalue k corresponds. A fundamental theorem in operator analysis states that we can express an operator in terms of its eigenfunctions and eigenvalues, both of which depend on the function space. Of particular importance is the so-called "Hilbert space," which has been carefully explored by theoretical physicists working in quantum mechanics. It contains "nice functions" such as x or $\sin x$, but not the singular, generalized functions that we shall need in order to introduce irreversibility into the statistical description. Every new theory in physics also requires new mathematical tools. Here, the basic novelty is our need to go beyond Hilbert space for unstable dynamical systems.

After these initial considerations, let us once again return to the Bernoulli map, where we can easily derive the explicit form of the evolution operator U, thereby obtaining $\rho_{n+1}(x) = U \rho_n(x) = \frac{1}{2}[\rho_n(\frac{x}{2}) + \rho_n(\frac{x+1}{2})]$. This equation means that after $(n + 1)$ iterations, the probability $\rho_{n+1}(x)$ at point x is determined by the values of $\rho_n(x)$ at points $\frac{x}{2}$ and $\frac{1+x}{2}$. As a consequence of the form of U, if ρ_n is a constant equal to α, ρ_{n+1} is also equal to α, since $U\alpha = \alpha$. The uniform distribution $\rho = \alpha$, which corresponds to equilibrium, is the distribution function reached through iteration of the shift, for $n \to \infty$.

On the contrary, if $\rho_n(x) = x$, we obtain $\rho_{n+1}(x) = \frac{1}{4} + \frac{x}{2}$. In other words, $Ux = \frac{1}{4} + \frac{x}{2}$ where the operator U transforms the function x into a different function, $\frac{1}{4} + \frac{x}{2}$. But we can easily find the eigenfunctions as defined above, in which the operator reproduces the same function multiplied by a constant. In the example $U(x - \frac{1}{2}) = \frac{1}{2}(x - \frac{1}{2})$, the eigenfunction is therefore $x - \frac{1}{2}$ and the eigenvalue $\frac{1}{2}$. If we repeat the Bernoulli map n times, we obtain $U^n(x - \frac{1}{2}) = (\frac{1}{2})^n(x - \frac{1}{2})$, which moves toward 0 for $n \to \infty$. The contribution $(x - \frac{1}{2})$ to $\rho(x)$ is therefore rapidly damped at a rate related to the Lyapunov exponent. The function $x - \frac{1}{2}$ belongs to a family of polynomials called the *Bernoulli polynomials*, denoted as $B_n(x)$, which are eigenfunctions of U with eigenvalues $(\frac{1}{2})^n$, where n is the degree of the polynomial.[3] When ρ is written as a superposition of Bernoulli polynomials, the polynomials of a higher degree disappear first because their damping factor is greater. This is the reason that the distribution function moves rapidly toward a constant. In the end, only $B_0(x) = 1$ survives.

We now need to express the distribution function ρ and the Perron-Frobenius operator U in terms of Bernoulli polynomials. Before we describe the result, however, we should once more emphasize the distinction between "nice" functions and "singular" functions (also called generalized functions or distributions, which are not to be confused with probability distributions), as it plays a crucial role. The simplest singular function is the delta function $\delta(x)$. As we saw in Chapter 1, Section III, $\delta(x - x_0)$ is zero for all values where $x \neq x_0$ and infinite where $x = x_0$. We have already noted that singular functions have to be used in conjunction with nice functions. For example, if $f(x)$ is a nice continuous function, the integral $\int dx \, f(x)\delta(x - x_0) = f(x_0)$ has a well-defined meaning. In contrast, the integral containing a product of singular functions, such as $\int dx \, \delta(x - x_0)\delta(x - x_0) = \delta(0) = \infty$, diverges and is therefore meaningless.

Our basic mathematical problem is defining the operator U in terms of its eigenfunctions and eigenvalues. This is called the spectral representation of the operator U. Once we have this representation, we can use it to express $U\rho$, that is, the effect of the Perron-Frobenius operator on the probability distribution ρ. Here we find a quite remarkable situation characteristic of deterministic chaos. We have already found a set of eigenfunctions, $B_n(x)$, the Bernoulli polynomials, which are nice functions, but there is a second set, $\tilde{B}_n(x)$, which is formed by singular functions related to derivatives of the δ-function.[4] To obtain the spectral representation of U and therefore $U\rho$, we need both sets of eigenfunctions. As a result, the statistical formulation for the Bernoulli map is applicable only to nice probability functions ρ and not to single trajectories that correspond to singular distribution functions represented by δ-functions. The spectral decomposition of U when applied to a δ-function contains products of singular functions that diverge and are meaningless. The equivalence between the individual description (in terms of trajectories represented by δ-functions) and the statistical description is broken. For continuous distribution ρ, however, we obtain consistent results that go beyond trajectory theory. We can calculate the rate of approach to equilibrium and therefore to an explicit dynamical formulation of irreversible processes that take place in the Bernoulli map. This outcome confirms the qualitative discussion in Chapter 1, Section III. Probability distribution takes into account the complex microstructure of the phase space. The description of deterministic chaos in terms of trajectories corresponds to an overidealization and is unable to express the approach to equilibrium.

Here we already encounter some of the most critical issues in modern mathematics. In fact, as we shall see in Chapters 5 and 6, the determination of eigenfunctions and eigenvalues is the central problem of statistical and quantum mechanics. The aim there, as well as for chaos, is to express an operator, such as U, in terms of its eigenfunctions and eigenvalues. When we succeed in doing so, we obtain the spectral representation of the operator. In quantum mechanics, such a representation has been achieved in simple situations in terms of nice functions. We may then use Hilbert space. The association between quantum mechanics and operator calculus in Hilbert space is so close that quantum mechanics is often considered an operator calculus in Hilbert space. In Chapter 6, we shall see that this is generally not the case.

Ultimately, to grasp the real world, we must leave Hilbert space. In the case of chaotic maps, we have to go out of Hilbert space because we need both the $B_n(x)$, which are nice functions, and the $\tilde{B}_n(x)$, which are singular functions. We can then speak of rigged Hilbert space, or Gelfand space. In more technical terms, we obtain an irreducible spectral representation of the Perron-Frobenius operator as it applies exclusively to nice probability distributions, and not to individual trajectories. These features are fundamental inasmuch as they are typical of unstable dynamical systems. We shall find them again in our generalization of classical dynamics in Chapter 5 and quantum mechanics in Chapter 6. The physical reasons for which we have to leave Hilbert space are related to the problem of persistent interactions mentioned above, which requires a holistic, nonlocal description. It is only outside Hilbert space that the equivalence between individual and statistical description is irrevocably broken, and irreversibility is incorporated into the laws of nature.

III

The Bernoulli map is not an invertible system. We mentioned earlier that an arrow of time already exists at the level of equations of motion. As our main problem is to describe the emergence of irreversibility in invertible dynamical systems, we shall now consider the baker map, or baker transformation, which is a generalization of the Bernoulli map. Let us take a square whose sides have length 1. First we flatten the square into a rectangle whose length is 2; then we cut it in half and build a new square. If we examine the lower part of the square, we see that after one iteration of this process (or mapping), it splits into two bands (see Figure 4.1). Moreover, the transformation is re-

Figure 4.1

The Baker Transformation

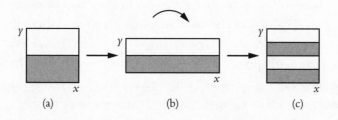

(a)　　　　　　　(b)　　　　　　　(c)

versible: The inverse transformation, which first reshapes the square into a rectangle with length $\frac{1}{2}$ and height 2, returns each point to its initial position.

For the Bernoulli map, the equations of motion are very simple: At each step, the coordinates (x,y) become $(2x, \frac{y}{2})$ for $0 \leq x < \frac{1}{2}$ and $(2x - 1, \frac{y+1}{2})$ for $\frac{1}{2} < x \leq 1$. To obtain the inverse baker transformation, we only have to permute x and y.

In the baker map, the two coordinates play different roles. The horizontal coordinate x is the dilating coordinate, which corresponds to the coordinate x in the Bernoulli map as it is multiplied by 2 (mod 1) at each mapping. The area of the square is preserved because we also have a contracting coordinate y; in the direction of the vertical coordinate, the points draw closer together while the square is being flattened into a rectangle. Since the distance between two points along the horizontal coordinate x doubles with each transformation, it will be multiplied by 2^n after n transformations. If we rewrite 2^n as $e^{n\log 2}$, as the number n of transformations measures time, the Lyapunov exponent is log 2, exactly as in the Bernoulli map considered in Section II. There is also a second Lyapunov exponent with the negative value −log 2, which corresponds to the contracting direction y.

The effect of successive iterations in the baker transformation is worthy of the same attention we gave to them in the discussion of the Bernoulli maps (see Figure 3.7). Here we start with points localized in a small portion of the square (see Figure 4.2), where we can clearly see the stretching effect of the positive Lyapunov exponent. As the coordinates x and y are limited to the interval $0 − 1$, the points are reinjected, leading to their uniform distribution throughout the square. By numerical simulation, we are also able to verify that if we start with the probability $\rho_n(x,y)$, the distribution moves rapidly toward unifor-

Figure 4.2

Numerical Simulation of the Baker Transformation

The maps are ordered according to the number of iterations, which represent time. (These numerical simulations are the work of Dean Driebe.)

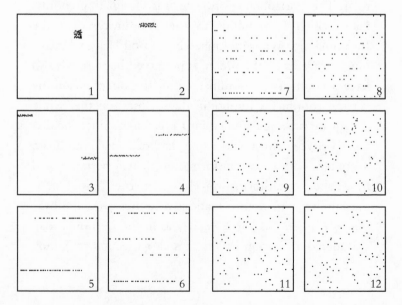

mity, as in the case of the Bernoulli shift (see Figure 3.8).

We can gain a great deal of insight into the mechanism of the baker transformation by representing it as a Bernoulli shift, as we did in Section I. Here we associate with each point (x,y) of the unit square the doubly infinite sequence of numbers $\{u_n\}$ defined by the binary representation

$$ x = \sum_{n=-\infty}^{0} \frac{1}{2^{n+1}} u_n, \; y = \sum_{n=1}^{\infty} \frac{1}{2^n} u_n, $$

where each u_n can take on the values 0 or 1. Each point x,y is represented by the series . . . $u_{-2}, u_{-1}, u_0, u_1, u_2$ . . . , in which . . . u_{-2}, u_{-1}, u_0 corresponds to the dilating coordinate x and $u_1, u_2, \ldots$ to the contracting coordinate y. For instance, the point $x = \frac{1}{4}$, $y = \frac{1}{4}$ will be represented by a series with $u_{-1} = 1$, $u_2 = 1$, with all other u_n being zero. By inserting these expressions into the equations of motion, we obtain the shift $u_n' = u_{n-1}$, which is again the Bernoulli shift. We see that the information contained in the initial conditions includes the entire past and future history of the system (Figure 4.3).

Successive iterations of the baker transformation lead to fragmentation of the shaded and unshaded areas, producing an increasing number of disconnected regions. Note that the digit u_0 determines whether the representative phase space point is in the left half ($u_0 = 0$) or the right half ($u_0 = 1$) of the unit square. Since the digits u_n, . . . can be determined by tossing a coin, the time iterates of u_n, $u_n' = u_{n-1}$, $u_n'' = u_{n-2}$ will have the same random properties. This shows that the process by which the point appears in the left or right half of the square can be considered a Bernoulli shift.

The baker transformation also shares an important property of all dynamical systems, known as recurrence.

Figure 4.3
Iterations of the Baker Transformation

Starting with the partition 0 (called the generating partition), we repeatedly apply the baker transformation. In moving toward the future, we generate horizontal bands. Similarly, by moving into the past, we generate vertical bands.

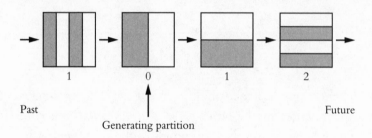

1   0   1   2

Past   Future

Generating partition

Consider a point (x,y) for which the sequence $\{u_n\}$ in the binary digit representation is finite or infinite but periodic, and x and y are then rational numbers. Since all u_n are shifted in the same way, every state of this kind will recycle identically after a certain period of time. The same holds true for most other states. To illustrate this concept, we shall consider the binary representation of an irrational point (x,y), which contains an infinity of nontrivial, nonrepeating digits. It can be shown that almost all irrationals contain a finite sequence of digits repeated an infinite number of times. Thus, a given sequence of $2m$ digits around position 0, which determines the state of the system to an error of 2^{-m}, will reappear an infinite number of times under the effect of the shift. Since m can be made as large as desired (although finite), almost every state will arbitrarily approach any point, including, of course, the initial position, an infinite number of times. In other words, most of the trajectories will traverse the entire phase space. This is the famous Poincaré recurrence theorem, which, together with time reversibility, was long advanced as an essential argument against the existence of genuinely dissipative processes. However, this view can no longer be sustained.

In summary, the baker transformation is *invertible, time reversible, deterministic, recurrent,* and *chaotic.* Demonstrating these properties through this example is especially useful, since these same properties characterize many real-world dynamical systems. As we shall see, despite these properties, chaos allows us to establish genuine irreversibility by setting up a description at the statistical level.

The dynamics of conservative systems involve laws of motion and initial conditions. Here the laws of motion are simple, but the concept of initial data demands a more detailed analysis. The initial conditions of a single trajectory correspond to an infinite set $\{u_n\}$ ($n = -\infty$ to $+\infty$). But in the real world, we can only look through a finite window. This means that we are able to control an arbitrary but limited number of digits u_n. Suppose that this window corresponds to $u_{-3}u_{-2}u_{-1}u_0 \cdot u_1u_2u_3$, all other digits being unknown (the dot indicates the separation between x and y digits). The Bernoulli shift $u'_n = u_{n-1}$ implies that at the next step, the previous series is replaced by $u_{-4}u_{-3}u_{-2}u_{-1} \cdot u_0u_1u_2$, which contains the unknown digit u_{-4}. More precisely, owing to the existence of a positive Lyapunov exponent, we need to know the initial position of the point with an accuracy of $N + n$ digits in order to be able to determine its position with an accuracy of N digits after n iterations.

As we saw in Chapter 1, the traditional means of solving this problem would be to introduce a coarse-grained probability distribution, which is not defined by single points, but rather by regions, as originally proposed by Paul and Tatiana Ehrenfest.[5] However, two points on an expanding manifold, even if not distinguishable by measurements of a given finite precision at time 0, will be separated, and thus observable, over time. Traditional coarse graining therefore cannot be applied to the dynamical evolution. This is one of the reasons for which we need a more sophisticated method.

First, however, we should analyze in more detail what the approach to equilibrium means in terms of the baker transformation.[6] In spite of the fact that this transformation is invertible, as are all dynamical systems, the evolutions for $t \to +\infty$ and $t \to -\infty$ are different. For $t \to +\infty$, we move toward increasingly narrow *horizontal* bands (see Figure 4.3). In contrast, for $t \to -\infty$, we move toward increasingly narrow *vertical* bands.

We see that for chaotic maps, dynamics lead to two types of evolutions. We thus obtain two independent descriptions, one characterizing the approach to equilibrium in our future (for $t \to +\infty$), and the other in our past (for $t \to -\infty$). Such dynamical decomposition is possible for both chaotic maps and nonintegrable classical and quantum systems, as we shall see later on. For a simple dynamical system, whether a harmonic oscillator or a two-body system, such decomposition does not exist; future and past cannot be distinguished. Which of the two descriptions for chaotic maps should we retain? We shall come back repeatedly to this question. For the moment, let us take into account the inherent universality that every irreversible process has in common. All arrows of time in nature have the same orientation: They all produce entropy in the same direction of time, which is by definition the future. We therefore have to retain the description corresponding to equilibrium reached in *our future*, that is, for $t \to +\infty$.

In Chapter 1, we mentioned the time paradox associated with the baker map: While the dynamics described by this map are time reversible, irreversible processes do appear at the statistical level. As in the Bernoulli map, we can introduce the Perron-Frobenius operator U defined by $\rho_{n+1}(x,y) = U \rho_n(x,y)$. But there is a fundamental difference. A general theorem states that for invertible dynamical systems there exists a spectral representation, defined on Hilbert space, which involves only nice functions.[7] Moreover, in this representation there is no damping, as the eigenvalues are modulo 1. Such a representation also exists for the baker transformation, but it is not of interest to us because it offers no new information regarding trajectories. We simply come back to $\delta(x - x_{n+1})\delta(y - y_{n+1}) = U \delta(x - x_n)\delta(y - y_n)$, a solution that is equivalent to the trajectory description.[8]

Exactly as we did for the Bernoulli map, we have to go out of Hilbert space to obtain additional information. For spectral representations in generalized space, which have recently been obtained, the eigenvalues are the same $(\frac{1}{2})^m$ as for the Bernoulli map.[9] Moreover, the eigenfunctions are singular functions, such as the $\tilde{B}_n(x)$ for the Bernoulli map. Again, these representations are irreducible in that they apply only to suitable test functions, obliging us to limit ourselves to continuous distribution functions. Single trajectories described by singular δ-functions are excluded. As is the case in the Bernoulli map, the equivalence between the individual description and the statistical description is broken. Only the statistical description includes the approach to equilibrium and therefore irreversibility.

For the baker map, there is one important new element involved, however, in comparison to the Bernoulli map:

The Perron-Frobenius equation can be applied to both future and past ($\rho_{n+1} = U \rho_n$ and $\rho_{n-1} = U^{-1} \rho_n$; here U^{-1} is the inverse of U). In the realm of Hilbert space spectral representations, this makes no difference because $U^{n_1 + n_2} = U^{n_1} U^{n_2}$, whatever the sign of n_1 and n_2 (remembering that the positive sign refers to the future, and the negative to the past). Hilbert space can be described as a *dynamical group*. In contrast, for irreducible spectral representations, there is an essential difference between future and past. The eigenvalues of U^n are expressed as $(\frac{1}{2}m)^n = e^{-n(m \log 2)}$. This formula corresponds to damping in the future ($n > 0$), and divergence in the past ($n < 0$). There now exist two different spectral representations—one for the future, and the other for the past. These two time directions, which are contained in the trajectory description (or Hilbert space), are now disentangled. The dynamical group is thereby broken into two *semigroups*. As previously mentioned, in accordance with our view that all irreversible processes are oriented in the same direction, we have to select the semigroup in which equilibrium is reached in our own future. Nature itself is described by a semigroup that distinguishes between past and future. There is an arrow of time. As a result, the traditional conflict between dynamics and thermodynamics is eliminated.

In summary, as long as we are considering trajectories, it seems paradoxical to speak of laws of chaos because we are dealing with the negative aspects of chaos, such as the exponential divergence of trajectories, which lead to uncomputability and apparent lawlessness. The situation changes drastically when we introduce the probabilistic description, which remains valid and computable at all times. It is therefore at the probabilistic level that the laws of dynamics have to be formulated for chaotic systems. In the simple examples studied above, irreversibility is linked only to Lyapunov time, but our research has recently been extended to more general maps that include such irreversible phenomena such as diffusion and various other transport processes.[10]

IV

As mentioned in Chapter 1, the success of the statistical description when applied to deterministic chaos stems from the fact that it takes into account the complex microstructure of phase space. In each finite region of phase space, there are exponentially diverging trajectories. The very definition of the Lyapunov exponent involves the comparison of neighboring trajectories. It is remarkable that irreversibility already emerges in simple situations involving only a few degrees of freedom. This is, of course, a blow to the anthropomorphic interpretation of irreversibility based on approximations that we ourselves are supposed to introduce. Unfortunately this interpretation, which was formulated after the defeat of Boltzmann, continues to be propagated today.

It is true that there is still a trajectory description if initial conditions are known with infinite precision. But this does not correspond to any realistic situation. Whenever we perform an experiment, whether by computer or some other means, we are dealing with situations in which the initial conditions are given with a finite precision and lead, for chaotic systems, to a breaking of time symmetry. Similarly, we could imagine infinite velocities, and therefore we would no longer need relativity theory, which is based on the existence of a maximum velocity—the velocity of light c in the vacuum—but the assumption of velocities greater than c corresponds to no known observable reality.

Maps are idealized models that cannot capture time's true continuity. As we now turn our attention to more realistic situations, of special importance to us will be nonintegrable Poincaré systems, where the break between the individual description (trajectories or wave functions) and the statistical description is even more striking. For these systems, the Laplace demon is powerless, whether his knowledge of the present is finite or infinite. The future is no more a given; it becomes a "construction," to use an expression of the French poet Paul Valéry.

Chapter 5

BEYOND NEWTON'S LAWS

I

Having analyzed maps that represent simplified models in Chapter 4, we come to the question at the very heart of our quest: What is the role of instability and persistent interactions in the framework of classical and quantum mechanics? Classical mechanics is the science upon which our belief in a deterministic, time-reversible description of nature is based. In responding to this question, we must first grapple with Newton's laws, the equations that have dominated theoretical physics for the past three centuries.

Quantum mechanics limits the validity of classical mechanics when applied to atoms and elementary particles. Relativity shows that classical mechanics also has to be modified when dealing with high energies or cosmology. Whatever the situation, we may introduce either an individual description (in terms of trajectories, wave functions, or fields) or a statistical description. Remarkably, at all levels, instability and nonintegrability break the equivalence of both descriptions. Consequently, we have to revise the formulation of the laws of physics in accordance with the open, evolving universe in which mankind lives.

As stated previously, our position is that classical mechanics is incomplete because it does not include irreversible processes associated with an increase in entropy. To include these processes in its formulation, we must incorporate instability and nonintegrability. Integrable systems are the exception. Starting with the three-body problem, most dynamical systems are nonintegrable. For integrable systems, the two modes of description—the trajectory description, based on Newton's laws, and the statistical description, based on ensembles—are equivalent. For nonintegrable systems, this is not so. Even in classical dynamics, then, we have to use the Gibbsian statistical approach (see Chapter 1, Section III). As we saw in Chapter 3, Section I, it is this approach that leads to the dynamical interpretation of equilibrium thermodynamics. It is therefore quite natural that we also have to employ the statistical description to include irreversible processes driving systems to equilibrium. In this way we can incorporate irreversibility into dynamics. As a result, there appear non-Newtonian contributions that can be consistently included in dynamics at the level of the statistical description. Moreover, these new contributions break time symmetry. We therefore obtain a probabilistic formulation of dynamics by means of which we can resolve the conflict between time-reversible dynamics and the time-oriented view of thermodynamics.

We are well aware that this step represents a radical departure from the past. Trajectories have always been considered primitive, fundamental tools of the trade. This is no longer the case. We shall encounter situations where trajectories "collapse," to borrow a term from quantum mechanics (see Section VII).

In hindsight, it is not surprising that we have had to abandon the trajectory description. As we saw in Chapter 1, nonintegrability is due to resonances, which express conditions that must be satisfied by frequencies. They are not local events that occur at given points in space and at a given instant in time. As such, they introduce elements that are quite foreign to the local trajectory description. Instead, we need a statistical description to formulate dynamics in situations where we expect irreversible processes and therefore an increase in entropy. Such situations, after all, are what we see in the world around us.

Indeterminism, as conceived by Whitehead, Bergson, and Popper, now appears in physics. This is no longer the result of some a priori metaphysical choice, but rather the need for a statistical description of unstable dynamical systems. Over the past decades, many scientists have proposed reformulations or extensions of quantum theory. But the fact that we now need to extend classical mechanics as well is quite unanticipated. Even more unexpected is the realization that this revision of classical mechanics can guide us in extending quantum theory.

II

Before we begin our revision of Newton's laws, let us summarize the fundamental concepts of classical mechanics. Consider the motion of a point of mass m. With the passage of time, its trajectory is described by its position, $r(t)$, its velocity, $v = dr/dt$, and its acceleration, $a = d^2r/dt^2$. Newton's basic equation relates acceleration a to force F through the formula $F = ma$. This formula includes the classical principle of inertia, that is, where there is no force, there is no acceleration, and the velocity remains constant. Newton's equation remains invariant when we shift from one observer to another who moves at a constant velocity with respect to the first. This is known as the Galilean invariance, which has been radically altered by relativity, as we shall see in Chapter 8. Here we are dealing with Newtonian, nonrelativistic physics.

We see that time takes its place in Newton's equation only by means of a second derivative. Newton's time, so to speak, is reversible, and future and past assume the same role. Moreover, Newton's law is deterministic.

Now consider a more general situation in which a system is formed by N particles. In three-dimensional space, we have the $3N$ coordinates $q_1, \ldots, q_{3N}$ and the corresponding velocities $v_1, \ldots, v_{3N}$. In modern formulations of dynamics, we usually define both the coordinates and velocities (or better, the momenta $p_1, \ldots, p_{3N}$, where in simple cases $p = mv$) as independent variables. As in Chapter 1, the state of the dynamical system is then associated with a point in phase space, and its motion with a trajectory in this space. The most important quantity in classical dynamics is the Hamiltonian H, which is defined as the energy of the system expressed in terms of the variables q and p. In general, H is the sum of the kinetic energy $E_{kin}(p)$ and the potential energy $V(q)$, where p and q signify the entire set of independent variables.

Once we have obtained the Hamiltonian $H(p, q)$, we can derive the equations of motion that determine the evolution of coordinates and momenta over the course of time. This procedure is familiar to all students of mechanics. Such equations, as derived from the Hamiltonian, are called the *canonical* equations of motion. Contrary to

Newton's equations, which are of the second order (that is, they contain the second time derivative), Hamiltonian equations are of the first order. For a single free particle, $H = \frac{p^2}{2m}$, the momentum p is constant over time, and the coordinate varies linearly when time $q = q_0 + \frac{p}{m}t$. By definition, for integrable systems, the Hamiltonian can be expressed only in terms of momenta (if necessary, after an appropriate change of variables). Poincaré studied Hamiltonians in the form $H = H_0(p) + \lambda V(q)$, which is the sum of an integrable contribution (the "free Hamiltonian" H_0) and a potential energy due to interactions (λ is a scaling factor that will be used later on). He showed that this class of Hamiltonians is generally not integrable, which is to say that we cannot eliminate interactions and go back to independent units. We already mentioned in Chapter 1 that nonintegrability is due to diverging denominators associated with Poincaré resonances, as a result of which we cannot solve the equations of motion (at least in powers of the coupling constant λ).

In the following pages, we shall concern ourselves primarily with nonintegrable large Poincaré systems (LPS). As we have seen, Poincaré resonances are associated with frequencies corresponding to various modes of motion. A frequency ω_k depends on the wavelength k. (Using light as an example, ultraviolet has a higher frequency ω and shorter wavelength k than infrared light.) When we consider nonintegrable systems in which the frequency varies continuously with the wavelength, we arrive at the very definition of LPS. This condition is met when the volume in which the system is located is great enough for surface effects to be ignored. This is why we call these systems large Poincaré systems.

A simple example of LPS would be the interaction between an oscillator with frequency ω_1 coupled with a given field. In this century of radio and television, we have all heard the term electromagnetic waves. The amplitude of these waves is defined by a field described by a function $\varphi(x,t)$ of position and time. As was established at the beginning of the century, a field can be thought of as the superposing of oscillations with frequencies ω_k whose wavelength k varies from the size of the system itself to the dimensions of elementary particles. In the oscillator-field interaction that we are considering, resonances appear each time a field frequency ω_k is equal to the oscillator frequency ω_1. When we try to solve the equations of motion of the oscillator in interaction with the field, we encounter Poincaré resonances $\frac{1}{(\omega_1 - \omega_k)}$, which correspond to divergences whenever $\omega_1 = \omega_k$. In other words, these terms tend toward infinity and therefore become meaningless. As we shall see, we can eliminate these divergences in our statistical description.

Poincaré resonances lead to a form of chaos. Indeed, innumerable computer simulations have shown that these resonances elicit the appearance of random trajectories, as is the case for deterministic chaos. In this sense, there is a close analogy between deterministic chaos and Poincaré nonintegrability.

III

As in previous chapters, we shall consider the probability distribution $\rho(q, p, t)$, whose evolution over time can easily be derived from the canonical equations of motion. We are now in the same situation as we were for chaotic maps, where we replaced the equations of motion with statistical descriptions associated with the Perron-Frobenius operator. In classical mechanics, we also encounter an evolution operator known as the Liouville operator L, which determines the evolution of ρ through the equation $i\frac{\partial \rho}{\partial t} = L\rho$. The time change of ρ is obtained by acting on ρ with the operator L. If the distribution function is time independent $\frac{\partial \rho}{\partial t} = 0$, then $L\rho = 0$. This corresponds to thermodynamic equilibrium. As we saw in Chapter 3, Section I, ρ then depends on only the energy (or the Hamiltonian), which is an invariant of motion.

The solution of dynamical problems at the statistical level requires determining the spectral representation of L, as was explained in Chapter 4 for chaotic systems. We therefore have to define its eigenfunctions and eigenvalues. We have seen that spectral representation depends on the functions which, as used in the past (and still appropriate for integrable systems), are in Hilbert space, the space of "nice" functions. According to a fundamental textbook theorem, operator L has real eigenvalues l_n in Hilbert space. In this case, evolution over time proves to be a superposition of oscillatory terms. In fact, the formal solution of the Liouville equation is $\rho(t) = \exp(-itL)\rho(0)$. The oscillatory term $\exp(-itl_n) = \cos tl_n - i \sin tl_n$ is associated with eigenvalue l_n, where future and past play the same role. In order to include irreversibility, we need complex eigenvalues such as $l_n = \omega_n - i\gamma_n$, which lead to exponential damping $e^{-\gamma_n t}$ for time evolution. This contribution progressively diminishes in the future ($t > 0$) but is increased in the past ($t < 0$), and thus time symmetry is broken.

However, obtaining complex eigenvalues is possible only when we leave Hilbert space. Our main objective is now to understand for which physical reasons we have to do so. This follows from the inescapable fact that there are *persistent* interactions in the natural world.[1] When we consider the room in which we sit, the molecules in the atmosphere are constantly colliding. This is quite different from *transitory* interactions, such as a finite number of molecules in a vacuum. The molecules then interact over a finite period of time, and eventually may escape into infinity. The distinction between persistent and transitory interactions takes on a crucial importance in moving from classical dynamics to thermodynamics. Classical dynamics extracts a given number of particles and considers their motion in isolation; irreversibility occurs when interactions never cease. In short, dynamics corresponds to a reductionist point of view in the sense that we consider a finite number of molecules in isolation. Irreversibility emerges from a more holistic approach in which we consider systems driven by a large number of particles as a whole. In making this distinction more precise, we shall indicate why we need singular distribution functions and must therefore leave Hilbert space.

IV

Transient interactions may be described by *localized* distribution functions. To describe persistent interactions in a large space such as the atmosphere, we need *delocalized* distribution functions. In defining more precisely the distinction between localized and delocalized distribution functions ρ, let us begin with a simple example. In a one-dimensional system, the coordinate x extends from $-\infty$ to $+\infty$. Localized distribution functions are concentrated on a finite section of the line. A special case is a single trajectory that is localized at a given point and moves along the line over the course of time. In contrast, delocalized distribution functions extend over the entire line. These two classes of functions describe various situations. As an example, let us consider *scattering*. In the usual scattering experiments, we prepare a beam of particles that we shoot at an obstacle (the scattering "center"). We then have the three stages represented in Figure 5.1.

In this experiment, the beam first approaches the scattering center, then interacts with it, and is finally in free motion again. The important point here is that the interaction process is transient. For delocalized distributions, on the other hand, the beam extends over the entire axis, and scattering neither starts nor stops. We then have what we call persistent scattering.

Transient scattering experiments have played a significant part in the history of physics by allowing us to study the interactions between elementary particles such as protons and electrons. Still, in many situations—particularly in macroscopic systems such as gases or liquids—we have persistent interactions because collisions never cease. In sum, transient interactions are related to localized distribution functions, such as trajectories, while persistent interactions are related to delocalized distributions, which extend over the entire system.

Thermodynamic systems are characterized by persistent

Figure 5.1

The Three Stages of Scattering

(a) The beam approaches the scattering center. (b) The beam intersects the scattering center. (c) The beam is once again in free motion.

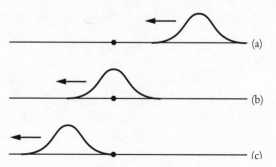

interactions, and must therefore be described by delocalized distributions. In defining these systems, we have to consider the *thermodynamic limit,* where the number of particles N and the volume V are increased, while their ratio, the concentration N/V, remains constant. Although formally we consider the limits $N \rightarrow \infty, V \rightarrow \infty$, there are, of course, no dynamical systems—not even the universe—where the number of particles is infinite. This limit simply means that surface effects described by the terms of $\frac{1}{N}$ or $\frac{1}{V}$ can be ignored. The thermodynamic limit plays a central role in all macroscopic physics. Without this concept, we could not even define states of matter such as gases, liquids, or solids, or describe the phase transitions between these states of matter. We would also be unable to distinguish between near-equilibrium and far-from-equilibrium situations, which were discussed in Chapter 2.

We shall now illustrate why the introduction of delocalized distribution functions forces us to leave the class of nice functions and therefore Hilbert space. In order to do so, we have to consider several elementary mathematical notions. In the first place, every student of mathematics is familiar with periodic functions such as $\sin \left(\frac{2\pi x}{\lambda} \right)$. This function remains invariant when we add to the coordinate x the wavelength λ, as

$$\sin \frac{2\pi x}{\lambda} = \sin \frac{2\pi(x + \lambda)}{\lambda}.$$

Other periodic functions are $\cos \frac{2\pi x}{\lambda}$, or the more complex combination

$$e^{i \frac{2\pi x}{\lambda}} = \cos \frac{2\pi x}{\lambda} + i \sin \frac{2\pi x}{\lambda}$$

Instead of the wavelength λ, we often use the wave vector $k = \frac{2\pi}{\lambda}$. The exponential e^{ikx} is called a plane wave.

In the second place, the classical theory of Fourier series

(or Fourier integrals) demonstrates that a function of the coordinate x, which we shall call $f(x)$, can be expressed as a superposition of periodic functions corresponding to wave vectors k, or more specifically, as a superposition of plane waves e^{ikx}. In this superposition, each plane wave is multiplied by an amplitude $\varphi(k)$, which is a function of k. This function $\varphi(k)$ is known as the Fourier transform of $f(x)$.

In short, we can go from a function $f(x)$ of coordinate x to a description $\varphi(k)$ in wave vectors k. Of course, the inverse transformation is equally possible. It is also important to note that there is a kind of duality between $f(x)$ and $\varphi(k)$. If $f(x)$ extends over a spatial interval Δx (and vanishes outside), $\varphi(k)$ extends over the "spectral" interval $\Delta k \sim \frac{1}{\Delta x}$. When the spatial interval Δx increases, the spectral interval Δk decreases, and vice versa.[2]

In Chapter 1, Section III and Chapter 3, Section II, we defined the singular function $\delta(x)$. As we saw, $\delta(x)$ differs from zero only at $x = 0$. The spatial interval Δx is therefore zero, and when $\Delta k \sim \frac{1}{\Delta x}$, the spectral interval is infinite. Inversely, delocalized functions for which $\Delta x \to \infty$ leads to singular functions in k such as $\delta(k)$. Thus, delocalized distribution functions are an essential element in describing persistent interactions. At equilibrium, the distribution function ρ is a function of the Hamiltonian H (see Chapter 3, Section I). The Hamiltonian contains the kinetic energy that is a function of the momenta p and not of the coordinates, and thus includes a delocalized part that has a singular Fourier transform. It is hardly astonishing that singular functions play a critical role in our dynamical description. Indeed, it is our need for these functions that forces us to leave Hilbert space. *Equilibrium* distributions that are functions of the Hamiltonian are already outside Hilbert space.

V

Let us now compare the trajectory description with the statistical description in terms of the Liouville operator (see Section III). Here we are in for quite a surprise because the statistical description introduces completely different concepts. This is obvious even in the simplest case where we consider the motion of a free particle along a line. As we saw in Section II, the coordinate q of the particle varies linearly over time, while the momentum p remains constant. On the contrary, the statistical description is defined in terms of the wave vectors k, associated with the Fourier transform of q, and the momentum p. We are used to dealing with wave vectors when we study acoustical or optical problems, but here wave vectors appear in a problem of dynamics. The reason is that for a free particle, the Liouville operator L is simply a derivative operator, where $L = \frac{ip}{m}\frac{\partial}{\partial x}$. As we noted in Chapter 4, Section I, the eigenfunctions are then exponentials exp (ikx) and the eigenvalues $\frac{pk}{m}$. The eigenfunction exp (ikx) is a periodic function, or plane wave, since exp $(ikx) = \cos kx + i \sin kx$. It extends over the entire space, in striking contrast with a trajectory localized at a single point. The solution of the equation of motion for a free particle is obtained in the statistical description through a superposition of plane waves. Of course, in this simple example, the two descriptions are expected to be equivalent. Using the theory of Fourier transformation, we can reconstruct the trajectory starting with plane waves (see Figure 5.2). Because the trajectory is concentrated at one point, we have to superpose plane waves extending over the entire length of the spectral interval ($\Delta k \rightarrow \infty$).

As a result, for $q = q_0$, the amplitudes of the plane waves

Figure 5.2

Superposition of Plane Waves

Trajectories resulting from the superposition of plane waves through constructive interference lead to a function characterized by dramatic peaking around $q = 0$.

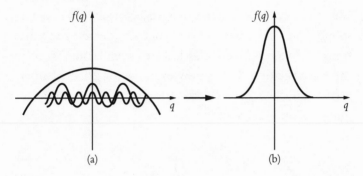

(a)   (b)

increase through *constructive* interference, while for $q \neq q_0$, they vanish through *destructive* interference. In integrable systems, the wave vector k is constant over time. By superposing the plane waves, we can reconstruct trajectories at any moment. But the important point to consider here is that the trajectory is no longer a primitive concept, but rather a derived concept as a construct of plane waves. It is thus conceivable that resonances may threaten the constructive interferences leading to a trajectory. This could not be considered as long as the trajectory was treated as a primitive, irreducible concept. Given that a trajectory is represented by a point in phase space, we can see that the collapse of trajectories would correspond to a situation in which a point decomposes over time into a multiplicity of points, exactly as in the diffusion process we analyzed in Chapter 1. The same initial condition would then lead to a multiplicity of trajectories, as was also the case in the diffusion process.

The eigenvalues $\frac{kp}{m}$ of the Liouville operator correspond to the frequencies appearing in Poincaré resonances. They depend on both k and p, and not on the coordinates. The use of wave vector k is therefore a logical starting point for discussing the role of these resonances. By using plane waves, we can describe not only trajectories (which correspond to transient interactions), but also delocalized situations. As we have seen, this leads to singular functions in the wave vector k. Let us now examine the effect of interactions on the statistical description by employing the language of wave vectors.

VI

Suppose that the potential energy V in the Hamiltonian is the sum of binary interactions. It then follows from well-established theorems that interactions between particles j and n modify the two wave vectors k_j and k_n, while their sum is conserved, giving us the conservation law $k_j + k_n = k'_j + k'_n$, where k'_j and k'_n are the wave vectors after interaction.[3]

We are able to describe dynamical evolution within the statistical formalism pictorially by considering a succession of events separated by free motion. At each event, the wave vectors k and momenta p are modified; between the events, they remain constant. Let us now examine the nature of these events in more detail.

In Chapter 3, Section I, we introduced the notion of correlations, which we shall now define with greater precision. The distribution function $\rho(q, p, t)$ depends on both coordinates and momenta. If we integrate this function over the coordinates, we lose all information about the position of particles, and thus correlations, in space. We obtain a function $\rho_0(p, t)$, which offers information only about momenta. For this reason, ρ_0 is known as the *vacuum of correlations*. On the other hand, by integrating over all coordinates except the coordinates q_i, q_j of particles i and j, we retain the information about possible correlations between particles i and j. This function, ρ_2, is called a binary correlation. We can define ternary correlations and beyond in a similar way. In the statistical description, it is important to replace the coordinates, which depend on the distribution functions through their Fourier transform, with wave vectors as they appear in the spectral decomposition of the Liouville operator.

We shall now take into account the law of conservation of wave vectors, in which each event is represented by a point, with two entry lines, k_j, k_n, and two exit lines, k'_j, k'_n, where $k_j + k_n = k'_j + k'_n$. Moreover, at each point, the momenta p of the interacting particles are modified, and a derivative operator $\frac{\partial}{\partial p}$ appears. The simplest event of this kind is illustrated in Figure 5.3.

We call the diagram in Figure 5.3 a propagation event, or propagation diagram. This corresponds to a modification of the binary correlation ρ_2 between particles j and n. But we can also start from the vacuum of correlations ρ_0, in which $k_j = k_n = 0$, and produce a binary correlation $\rho_{k_j k_n}$, with $k_j + k_n = 0$ to conserve the sum of the wave vec-

Figure 5.3

Propagation Diagram

A dynamical event corresponding to the interaction of two particles leads from wave vectors k_j, k_n to k'_j, k'_n.

Figure 5.4
Creation Fragment

A dynamical event transforms the vacuum of correlations into a binary correlation $l, -l$.

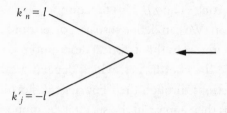

tors (see Figure 5.4). We then have what is known as a creation of correlation diagram, or creation fragment. We also have destruction fragments, as presented in Figure 5.5, which transform binary correlations into the vacuum of correlations.[4]

We now begin to see dynamics as a *history of correlations*. Figure 5.6 represents, for example, the emergence of a five-particle correlation starting from the vacuum of correlations. Events associated with interactions produce correlations.

We can now introduce the effect of Poincaré resonances into the statistical description of dynamics. These resonances couple dynamical processes exactly as they couple

Figure 5.5
Destruction Fragment

A dynamical event transforms the binary correlation $l, -l$ into the vacuum of correlations.

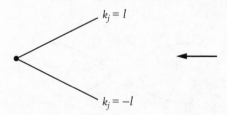

Figure 5.6

Evolution of Correlations

The four events at points 0_1, 0_2, 0_3, 0_4 transform the vacuum of correlations into a five-particle correlation.

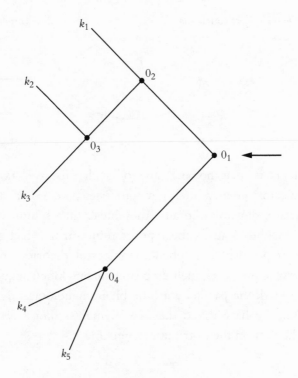

harmonics in music. In our description, they couple creation and destruction fragments (see Figure 5.7), which leads to *new* dynamical processes that start from a given state of correlations (of which the vacuum of correlations is merely one example) and eventually return to exactly the *same* state. In Figure 5.7, these dynamical processes are depicted as bubbles. While the state of correlations is preserved, the distribution of momenta is changed (remembering that each vortex introduces a derivative operator $\frac{\partial}{\partial p}$).

These bubbles correspond to events that must be considered *as a whole*. They introduce *non–Newtonian* ele-

Figure 5.7

Bubble Due to Poincaré Resonances

Poincaré resonances couple the creation and destruction of correlations, and lead to diffusion.

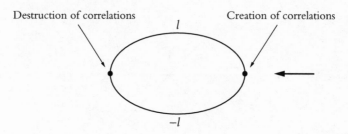

Destruction of correlations   *l*   Creation of correlations

−l

ments in that no analogue of such processes exists in trajectory theory. Such new processes have a dramatic effect on dynamics because they break time symmetry. Indeed, they lead to the type of diffusion that had always been postulated in phenomenological theories of irreversible processes, including Boltzmann's kinetic equation. To mark the parallel with the phenomenological description, we have called the new elements *collision operators*. They act on the distribution functions.*

---

*We saw in Chapter 1, Section III that Poincaré resonances between frequencies lead to divergences with small denominators. Here the frequency of a particle of momentum p is kp/m, where k is the wave vector (see Section IV). For LPS, in which k is a continuous variable, we can avoid the divergences and express the resonances in terms of δ-functions. This involves a branch of mathematics associated with analytical continuation (see the references in the chapter notes). For a two-body process, the argument of the δ-function is k/m $(p_1 - p_2)$, leading to contributions whenever the frequencies kp_1/m and kp_2/m are equal, and otherwise vanishing. The wave vector $k = 0$ therefore plays an especially important role wherein the argument of the δ-function vanishes, remembering that $\delta(x) = \infty$ for $x = 0$ and $\delta(x) = 0$ for $x \neq 0$. A vanishing wave vector k corresponds to an infinite wavelength, and thus to a process that is *delocalized in space*. Hence, Poincaré resonances cannot be included in the trajectory description.

Our approach includes the usual kinetic theory, but only as a special case. Traditionally this theory, as introduced by Maxwell, was centered around the evolution of the velocity distribution, where it appeared that only a few collisions would be sufficient to reestablish equilibrium if disturbed at the initial time. Our approach, on the contrary, takes into account the progressive buildup of higher and higher correlations involving more and more particles. This process requires long time scales, in agreement with the numerical simulations that have been available for many years.[5] As a result, irreversibility leads to long memory effects that profoundly alter macroscopic physics.[6]

Many new results that go beyond the traditional kinetic theory have already been obtained. However, it is outside the scope of this book to describe them. They will be covered in greater detail in a separate monograph.[7]

Suffice it to say that we are beginning to understand what irreversibility really means. Let us consider the simple analogy of the aging process. On our time scale, the atoms that make up our bodies are immortal. What is changing is the relation between the atoms and molecules. In this sense, aging is a property of populations, and not individuals. This is also true of the inanimate world.

VII

Let us now return to our original objective, which is the solution of the dynamical problem at the statistical level in terms of the distribution function ρ. As was the case for deterministic chaos, this solution involves the spectral representation of the evolution operator, which in classical dynamics is the Liouville operator. First we consider delocalized distribution functions associated with persistent interactions that lead to singular functions (see Sections III and IV). As a result, we have to leave Hilbert space, which is limited to localized nice functions. We then introduce Poincaré resonances, which, as we saw in Section VI, lead to new dynamical processes connected with diffusion.

Once we have included these two features, we obtain an irreducible, complex spectral representation. Again, *complex* means time symmetry is broken, and *irreducible* means we cannot return to a trajectory description. The laws of dynamics now take on new meaning. By incorporating irreversibility they express not certitudes but possibilities. Only if we relax our conditions and consider localized distribution functions associated with a finite number of particles can we recover the Newtonian trajectory description. But in general diffusion processes dominate.

There are therefore many situations in which we can expect deviations from Newtonian physics, and where our predictions have already been verified by extensive computer simulations. In Section IV, we introduced the thermodynamic limit, where the number of particles $N \to \infty$ and volume $V \to \infty$, while $\frac{N}{V}$ = the concentration that remains constant. In this limit, interactions go on forever, and only the statistical description applies. It has been shown by extensive numerical simulations that even if we start with a trajectory involving an ever-increasing number of particles, diffusive processes take over, and the trajectory "collapses" because it is transformed over time into a delocalized singular distribution function.[8]

Our new kinetic theory is of great interest in describing dissipative processes for all time scales, as observed in the laboratory or the ecosphere. But this is only one of its many novel features. Because of Poincaré resonances, the dynamical processes described in this section lead to long-range correlations, even if the forces between the particles are short range. The only exception is the state of equilibrium, where the range of correlations is determined by that of the forces between the particles. This explains the fact, as stated in Chapter 2, that nonequilibrium allows for a new coherence, which is clearly manifested by chemical oscillations and hydrodynamic flows. We now recognize that equilibrium physics gave us a false image of matter. Once again, we are faced with the fact that matter in equilibrium is "blind," while in nonequilibrium it begins to "see."

In sum, we are now able to go beyond Newtonian mechanics. The validity of the trajectory description used in classical mechanics is severely limited. Thermodynamics is incompatible with trajectory description, as it requires a statistical approach both at equilibrium and out of equilibrium. The fact that the vast majority of the dynamical systems corresponding to the phenomena that surround us are LPS is the reason why thermodynamics is universally valid. Transient dynamical interactions such as *scattering* are not representative of the situations that we encounter in the natural world, where interactions are persistent. The collision processes that appear in our statistical description as a result of Poincaré resonances are essential in that they break time symmetry and lead to evolutionary patterns in accordance with the thermodynamic description.

The microscopic depiction of nature associated with thermodynamics has little to do with the comfortable time-symmetrical description scientists have traditionally taken from Newtonian principles. Ours is a fluctuating, noisy, chaotic world more akin to what the Greek atomists imagined. In Chapter 1, we described Epicurus' dilemma. The clinamen he envisaged no longer belongs to a philosophical dream that is foreign to physics. It is the very expression of dynamical instability.

Of course, dynamical instability provides only those conditions necessary to generate evolutionary patterns of nature. Once we have achieved our statistical description, we can also formulate the additional factors we need in order to observe the emergence of complexity—of dissipative structures at the macroscopic level. We now begin to understand the dynamical roots of organization, the dynamics at the root of complexity that are essential for self-organization and the emergence of life.

Chapter 6

A UNIFIED FORMULATION
OF QUANTUM THEORY

I

There are fundamental differences between classical Newtonian dynamics and quantum theory. But in both cases there exist an individual description in terms of trajectories or wave functions (see Chapter 1, Section IV) and a statistical description in terms of probability distributions. As we have already seen, Poincaré resonances appear in classical as well as quantum theory. We can therefore anticipate that the results obtained in classical mechanics will also apply to quantum theory. In fact, in both instances we have achieved a new statistical formulation applicable to LPS outside Hilbert space. This description includes time-symmetry breaking, and is irreducible to the individual description in terms of quantum wave functions.

In spite of quantum theory's astonishing success, discussions about its conceptual foundations have not abated. After seventy years, they are as lively as ever.

For example, in his recent book *Shadows of the Mind,* Roger Penrose distinguishes between "Z mysteries" (for quantum *puzzles*) and "X mysteries"(for the quantum *paradox*) in quantum behavior.[1] Furthermore, the role of nonlocality seems intensely problematic. Given that locality is a property associated with the Newtonian pointwise trajectory description, it is not surprising that quantum theory, which includes the wave aspect of matter, leads to a form of nonlocality.[2]

The "collapse" of the wave function, which seems to require a dualistic formulation of quantum theory, represents a further complication. On the one hand, we have the basic Schrödinger equation for wave functions, which is time reversible and deterministic, exactly as is Newton's equation; on the other, we have the measurement process associated with irreversibility and the collapse of the wave function. This dualistic structure is the basis of John von Neumann's argument in his famous book, *Mathematical Foundations of Quantum Mechanics.*[3] This situation is indeed bizarre because in addition to the basic time-reversible, deterministic Schrödinger equation, there would be a second dynamical law associated with the collapse (or reduction) of the wave function. Until now, however, no one has been able to describe the link between these two laws of quantum theory, nor has anyone succeeded in giving a realistic interpretation of the reduction of the wave function. This is the *quantum paradox.*

The quantum paradox, which derives from the dualistic structure of quantum theory, is closely related to another problem. Our conclusion is that quantum theory is incomplete. Like classical trajectory theory, it is time symmetric, and therefore cannot describe irreversible processes such as the approach to thermodynamic equilibrium. This is particularly curious because quantum theory began in 1900 with Max Planck's successful description of black body radiation in equilibrium with matter. Even today, in spite of the great advances made by Albert Einstein and Paul A. M. Dirac, we still have no exact quantum theory describing the approach to equilibrium when radiation interacts with matter. (As we shall see, this is related to the fact that quantum theory describes integrable systems. We shall come back to this challenge in Section IV.) We need both equilibrium and nonequilibrium physics to describe the world around us. An example of an equilibrium situation is the famous residual black body radiation at 3°K, which originated at a time close to the big bang. A large part of macroscopic physics also deals with equilibrium systems, whether they are solids, liquids, or gases. There is thus a gap between quantum theory and thermodynamics as deep as that between classical theory and thermodynamics. Remarkably, the same method employed in extending classical mechanics in Chapter 5 also permits us to unify quantum theory and thermodynamics. Indeed, our approach eliminates the dualistic structure of quantum mechanics, and thus eliminates the quantum paradox. We arrive at a realistic interpretation of quantum theory because the transition from wave functions to ensembles can now be understood as the result of Poincaré resonances without the mysterious intervention of an "observer" or the introduction of other uncontrollable assumptions. In contrast to other attempts to extend quantum theory, as noted in Chapter 1, our own approach makes well-defined predictions that are testable. Thus far, they have been confirmed by every numerical simulation performed.[4]

Our thinking constitutes a return to realism, but emphatically not a return to determinism. On the contrary, we move even farther away from the deterministic vision of classical physics. We agree with Popper when he writes, "My own point of view is that indeterminism is compatible with realism, and that the acceptance of this fact allows us to adopt a coherent objective epistemology of the whole of quantum theory, and an objectivist interpretation of probability." We shall therefore endeavor to bring into the realm of physics what Popper called his metaphysical dream: "It is likely that the world would be just as indeterministic as it is even if there were no observing subjects to experiment with it, and to interfere with it."[5] Thus we will show that the quantum theory of unstable dynamical systems with persistent interactions leads, as in classical systems, to a description that is both statistical and realistic. In this new formulation, the basic quantity is no longer the wave function corresponding to a probability *amplitude,* but *probability itself.* As in classical physics, probability emerges from quantum mechanics as a fundamental concept. In this sense, we are on the eve of the triumph of the "probabilistic revolution," which has been going on for centuries. Probability is no longer a state of mind due to our ignorance, but the result of the laws of nature.

II

The observation that the interaction between atoms and light leads to well-defined absorption and emission frequencies was the starting point for the formulation of quantum mechanics. The atom was described by Niels H. D. Bohr in terms of discrete energy levels. In accordance with experimental data (the Ritz–Rydberg principle), the frequency of spectral lines is the *difference between two energy levels.* Once these levels are known, we can predict the frequency of spectral lines. The problems of spectroscopy can be reduced to the calculation of levels of energy. But how can we reconcile the existence of well-defined energy levels, which decisively influenced the history of quantum theory, with the Hamiltonian concept that is so important to classical theory? The classical Hamiltonian expresses the energy of a dynamical system in terms of coordinates q and momenta p, and therefore takes on a continuous set of values. It cannot lead to discrete energy levels. For this reason, the Hamiltonian H is replaced in quantum theory by the Hamiltonian *operator* H_{op}.

We have repeatedly used operator formalism (the Perron-Frobenius operator was introduced in Chapter 4, and the Liouville operator in Chapter 5), but it was in quantum theory that operator calculus was first introduced into physics. In the situations studied in Chapters 4 and 5, we needed operators to achieve the statistical description. Here, even the individual level of description corresponding to wave functions requires operator formalism.

The basic problem in quantum mechanics is the determination of the eigenfunctions u_α and the eigenvalues E_α of the Hamilton operator H (we shall omit the subscript *op* wherever possible). The eigenvalues E_α, which are identified with the *observed values* of the energy levels, form the spectrum of H. We speak of a *discrete spectrum* when successive eigenvalues are separated by finite distances. If the spacing between levels tends toward zero, we then speak of a *continuous spectrum*. For a free particle in a one-dimensional box with a length of L, the spacing of the energy level is inversely proportional to L^2. As a consequence, when $L \to \infty$, this spacing moves toward zero, and we obtain a continuous spectrum. By definition, the word "large" in large Poincaré systems (LPS) means precisely that these systems have a continuous spectrum. As in classical theory, the Hamiltonian is here a function of coordinates and momenta. However, because the Hamiltonian is now an operator, these quantities, and therefore all dynamical variables, now have to be treated also as operators. For today's physicists, the transition from functions to operators that takes place in quantum theory seems perfectly natural. They now manipulate operators with the ease with which most of us manipulate natural numbers. Nonetheless, for classical physicists such as the great Dutch scientist Hendrik Antoon Lorentz, the introduction of operators was barely acceptable, and even repulsive. In any case, individuals such as Werner Heisenberg, Max Born, Pascual Jordan, Erwin Schrödinger, and Paul Dirac, who daringly introduced operator formalism into physics, deserve our admiration. They drastically changed our description of nature in defining the conceptual difference between a physical quantity (represented by an operator) and the numerical values this physical quantity may take on (the eigenvalues of the corresponding operator). This radical change in outlook has had far-reaching and profound implications for our conception of reality.

As an example of the sophistication of operator formalism, consider the commutation relations between two operators. These operators commute if the order of their application to a function is immaterial. They do not commute if the order of their application changes the result. For instance, multiplying a function $f(x)$ by x and then differentiating it with respect to x does not lead to the same result as first differentiating $f(x)$ and then multiplying it by x. This can easily be verified. Operators that do not commute exhibit different eigenfunctions; if they do commute, they have common eigenfunctions.

The famous Heisenberg *uncertainty principle* follows from the fact that the coordinate and momentum operators, as defined in quantum theory, do not commute. In all textbooks on quantum mechanics, it is shown that in the "coordinate representation," the operator corresponding to a coordinate q_{op} has eigenvalues that are the coordinates of the quantum object. The operator q_{op} may therefore be identified with the classical coordinate q. In contrast, the momentum operator p_{op} is defined by the derivative operator $\frac{h}{i}\frac{\partial}{\partial q}$ which is a derivative in respect to q. The two operators q_{op} and p_{op} thus do not commute, and have no common eigenfunctions.[6] In quantum mechanics, we may use various representations. In addition to the coordinate representation, we have the momentum representation, where the momentum operator is simply p, and coordinates are represented by derivative operators $\frac{h}{i}\frac{\partial}{\partial p}$. Whatever the representation, the two operators do not commute.

The fact that q_{op} and p_{op} do not commute means that we cannot define states of a quantum object for which both the coordinate and the momentum take on well-defined values. This is the root of Heisenberg's uncertainty reaction, which forces us to abandon the "naive realism" of classical physics. We are able to measure the momentum or the coordinate of a given particle, but we cannot say that this particle has well-defined values for both its momentum and its coordinates. This conclusion was reached sixty years ago by Heisenberg and Born, among others. Even so, discussions about the meaning of uncertainty relations still go on, and some scientists have not yet given up the hope of restoring the traditional deterministic realism of classical mechanics.[7] This was one of the reasons for Einstein's dissatisfaction with quantum theory. We should note that

Heisenberg's uncertainty principle is compatible with a deterministic time-symmetrical description of nature (the Schrödinger equation).

What do we mean when we say that a quantum system is in a particular "state"? In classical mechanics, the state is a point in phase space. Here it is described by a wave function whose evolution over time is expressed by the Schrödinger equation $ih/2\pi\ \partial\Psi(t)/\partial t = H_{op}\Psi(t)$. This equation identifies the time derivative of the wave function Ψ with the action of the Hamiltonian operator on Ψ. It is not derived, but rather is assumed at the start, and can thus be validated only by experiment. It is the fundamental law of nature in quantum theory.* Note the formal analogy with the Liouville equation in Chapter 5, Section III, where the basic difference is that L (the Liouville operator) acts on distribution functions ρ, while H_{op} acts on wave functions.

We have already mentioned that a wave function corresponds to a probability amplitude. The parallel that guided Erwin Schrödinger in formulating his equation was that of classical optics. In contrast to the trajectory equations of classical mechanics, the Schrödinger equation is a wave equation. It is a *partial* differential equation because in addition to the time derivative, there are also derivatives with respect to coordinates appearing in H_{op} (remembering that in the coordinate representation, the momentum operator is a derivative with respect to coordinates). But classical and quantum equations have an essential element in common: They both correspond to a deterministic description.

---

*There are various extensions of the Schrödinger equation and the relativistic Dirac equation, but they are not necessary to this discussion.

Once Ψ is known at some arbitrary time t_0, together with appropriate boundary conditions (such as $\Psi \to 0$ at infinite distances), we may calculate Ψ for arbitrary times in the future as well as in the past. In this sense, we reinstitute the deterministic view of classical mechanics, but it now applies to wave functions, and not trajectories.

As in the classical equations of motion, the Schrödinger equation is time reversible. When we replace t by $-t$, the equation remains valid. We only have to replace Ψ with its complex conjugate Ψ^*. As a consequence, if we observe the transition of Ψ from Ψ_1 to Ψ_2 at time t_2, where $t_2 > t_1$, we can also observe a transition from ψ^*_2 to ψ^*_1. It is worth reminding ourselves of Arthur Stanley Eddington's remark at an early stage in quantum mechanics to the effect that quantum probabilities are "obtained by introducing two symmetrical systems of waves traveling in opposite directions of time."[8] Indeed, as we have seen, the Schrödinger equation is a wave equation describing the evolution of probability amplitudes. If we now take the complex conjugate of the Schrödinger equation, that is, if we replace i by $-i$, Ψ by Ψ^* (supposing that H_{op} is real), and t by $-t$, we return to Schrödinger's equation. As stated by Eddington, Ψ^* may therefore be viewed as a wave function propagating into the past. Furthermore, as mentioned in Chapter 1, probability proper is obtained through multiplying Ψ by its complex conjugate Ψ^* (that is, $|\Psi|^2$). Since Ψ^* may be interpreted as Ψ evolving backward in time, the definition of probability implies the meeting of two times, one stemming from the past and the other the future. In quantum theory, probabilities are thus time symmetric.

We now see that in spite of their fundamental differences, both classical and quantum mechanics correspond to laws of nature that are deterministic and time reversible. No difference between past and future appears in these formulations. As we noted in Chapters 1 and 2, this leads to the time paradox. In quantum mechanics, it also leads to the quantum paradox, due to the need to introduce a dualistic formulation of quantum theory. In both classical theory and quantum theory, the Hamiltonian plays a central role. In quantum theory, its eigenvalues determine the energy levels, while, according to the Schrödinger equation, it also determines the time evolution of the wave function.

As in the preceding chapter, we shall concentrate on systems in which the Hamiltonian H is the sum of a free Hamiltonian H_0 and a term produced by the interactions λV, whereby $H = H_0 + \lambda V$. The time history of such systems can then be described by transitions between eigenstates of H_0 induced by these interactions.

As long as we remain in Hilbert space, the eigenvalues E_α of H are *real* (like the Liouville operator, H is also "hermitian," and hermitian operators have real eigenvalues in Hilbert space). The evolution of the wave function is a superposition of oscillating terms such as $\exp(-iE_\alpha t)$. There are, however, irreversible processes in quantum mechanics, such as the quantum leaps in Bohr's theory, where excited atoms decay through the emission of photons or unstable particles (see Figure 6.1) or the decay of unstable particles.

Can these processes be included in Hilbert space within the framework of traditional quantum theory? Decay processes occur in large systems. If an excited atom were kept in a cavity, the emitted electron would bounce back, and there would be no irreversible process. As we have seen, the time evolution of the wave function is described

Figure 6.1

Decay of an Excited Atom

The atom "falls" from the excited state to the ground state with the emission of a photon.

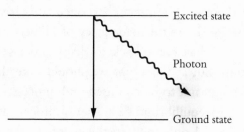

by a superposition, or sum, of oscillatory terms. With the limit of large systems, this sum becomes an integral, and acquires new properties. In the case of the decay of excited atoms as described by Figure 6.1, the probabilities $|\Psi|^2$ decay *almost* exponentially over time. Here the word *almost* is essential: As long as we remain in Hilbert space, there are deviations from the exponential for both very brief times (the same order as the frequency of oscillations of the electron around the nucleus ~ 10^{-16} seconds) and very long times (for example, ten to one hundred times the lifetime of an excited state, which is ~ 10^{-9}). However, in spite of a great number of experimental studies, no deviations from exponential behavior have yet been detected. This is indeed fortunate, because if they did exist, it would raise serious questions about the entire theoretical system of particle physics.

Suppose that we prepare a beam of unstable particles, let it decay, and later on prepare a second beam. Imagine the strange situation of the two beams prepared at different times having different decay laws. We could then distinguish between them just as we do between older and younger individuals! This fantasy would be a violation of the principle of indistinguishability for elementary particles, which has led to some of the greatest successes of quantum theory.* The precise exponential behavior observed thus far shows the inadequacy of Hilbert space description. We shall come back to decay processes in the next section, but at this point we should note that such processes ought not to be confused with processes driving the system to equilibrium. The decay process as represented in Fig. 6.1 only transfers the energy of the atom to the photons.

III

As we have seen, the main issue in quantum mechanics is the solution of the eigenvalue for the Hamiltonian. There are only a few quantum systems in which this problem has been solved exactly. In order to do so, we generally need to use a perturbational approach. As mentioned, we start with a Hamiltonian in the form $H = H_0 + \lambda V$, where H_0 corresponds to a Hamiltonian operator for which we have solved the eigenvalue (the "free" Hamiltonian) and V is a perturbation coupled with H_0, through the so-called coupling constant λ. We assume that we know the solution of the eigenvalue $H_0 u_n^{(0)} = E_n^{(0)} u_n^{(0)}$, and that we wish to solve the equation $H u_n = E_n u_n$. The standard procedure, which is Schrödinger's perturbational method, is to expand both

---

*These include the explanation of superfluidity and the quantum theory of solid state.

the eigenvalues and eigenfunctions in terms of powers of the coupling constant λ.

The perturbational approach leads to a recurrence procedure involving equations for each order in λ. The solution of these equations implies the use of terms such as $1/(E_n^{(0)} - E_m^{(0)})$, which become ill defined when the denominator vanishes. This situation again corresponds to resonances,* and once more we encounter the divergence problem that lies at the very center of Poincaré's definition of nonintegrable systems.

However, there is an essential difference here. We have already introduced the distinction between discrete and continuous spectrums. In quantum mechanics, this difference becomes crucial. In fact, when the spectrum is discrete, it is generally possible to avoid the divergence problem through an appropriate choice of the unperturbed Hamiltonian.† Since all finite quantum systems have a discrete spectrum, we can then conclude that they are integrable.

The situation changes dramatically when we turn to large quantum systems involving excited atoms, scattering systems, and so on. In this case, the spectrum is continuous, which brings us back to LPS. The example of a particle coupled with a field, which we presented in Chapter 5, Section V, also applies to quantum systems. We then have resonances whenever the frequency ω_l associated with the particle is equal to a frequency ω_k associated with the field.

---

*In quantum mechanics, to each energy E corresponds a frequency ω expressed as $E = (h/2\pi)\omega$.

†In more technical terms, we first raise the degeneracy by an appropriate transformation.

The only difference is that in quantum theory, frequencies are associated with energies. The eigenvalue E_α corresponds to the frequency $\frac{h}{2\pi}\omega_\alpha$, where h is Planck's constant.

The example in Figure 6.1, which corresponds to an LPS, illustrates that we have resonance each time the energy difference between the two levels is equal to the energy of the photon that is emitted.

As in the case of deterministic chaos studied in Chapter 4, we can extend the eigenvalue problem to singular functions outside Hilbert space. The formal solution of the Schrödinger equation is $\Psi(t) = U(t)\Psi(0)$, where $U(t) = e^{-iHt}$; $U(t)$ is the evolution operator that links the value of the wave function at time t to that at the initial time $t = 0$. Both future and past play the same role, since $U(t_1)U(t_2) = U(t_1 + t_2)$, whatever the sign of t_1 and t_2. This property defines what is called a dynamical group. Outside Hilbert space, the dynamical group splits into two semigroups. There are then two functions corresponding to the excited atom: The first, φ_1, decays exponentially in the future ($\varphi_1 \sim e^{-t/\tau}$), while the second, $\tilde{\varphi}_1$, decays in the past ($\tilde{\varphi}_1 \sim e^{t/\tau}$). Only one of these two semigroups is realized in nature. In both cases, there is an *exact* exponential decay (in contrast to the approximate one described in the preceding section). This was the first such example studied, notably by Arno Bohm and George Sudarshan, who showed that in order to obtain exact exponential laws and avoid the difficulties mentioned in Section II, Hilbert space must be abandoned.[9] However, in their approach, the central quantity remains the probability amplitude, and the basic paradox of quantum mechanics (the collapse of the wave function) is not solved. As already mentioned, the decay of excited atoms or unstable particles corresponds only to a transfer of energy from one system (the excited atom) to the other (the photon). The approach to equilibrium requires a fundamental modification of quantum theory. As in classical mechanics, we have to go from the individual description, associated with wave functions, to the statistical description, associated with ensembles.

IV

In the transition from the individual to the statistical description, quantum theory introduces certain specific features as compared to classical mechanics. There, as we saw in Chapter 5, the statistical distribution function is a function of both the coordinates and momenta. A trajectory corresponds to the delta function (see Chapter 1, Section III). In quantum mechanics, the quantum state, as associated with a wave function, is described by a continuous function of the independent variables. We can either take the coordinates as independent variables and consider $\Psi(q)$, or we can take the momenta and consider $\Psi(p)$. Heisenberg's uncertainty principle prevents us from taking both. The definition of a quantum state therefore involves only half of the variables that are used in the definition of the classical state.

The quantum state Ψ represents a probability *amplitude* for which the corresponding probability ρ is given by the product of the amplitudes $\Psi(q)$ and $\Psi^*(q')$, and is therefore a function of two sets of variables, q and q' or p and p'. We can thus write $\rho(q, q')$ or $\rho(p, p')$, where the first expression corresponds to the coordinate representation, and the second to the momentum representation, which will be especially useful to us. In quantum mechanics, the probability ρ is often called the "density matrix" (matrices, as studied in algebra, also have two indices). We can easily write the equation of evolution for ρ because the equation for Ψ (the Schrödinger equation) is already known. The evolution equation for ρ is the quantum Liouville equation, whose explicit form is $i\hbar\,(\frac{\partial\rho}{\partial t}) = H\rho - \rho H$, which is the "commutator" of ρ with H. This shows that when ρ is a function of H, we have an equilibrium situation. Then $\partial\rho/\partial t = 0$, as H commutes with a function of itself.

Now that we have considered the distribution function ρ, which corresponds to a single wave function, we can also consider situations in which ρ corresponds to a "mixture" of various wave functions. In both cases, the Liouville equation remains the same.

For integrable systems, the statistical formulation introduces no new features. Suppose that we know the eigenfunctions $\varphi_\alpha(p)$ and the eigenvalues E_α of H. The eigenfunctions of L are then the products $\varphi_\alpha(p)\varphi_\beta(p')$ and the eigenvalues the differences $E_\alpha - E_\beta$. The problems involved in deriving the spectral representations of H and L are thus equivalent.

The eigenvalues $E_\alpha - E_\beta$ of L correspond directly to the frequencies measured in spectroscopy, where the time evolution of the distribution function ρ is a superposition of oscillating terms $e^{-i(E_\alpha - E_\beta)t}$. Again, there is no approach to equilibrium. Moreover, for those situations in which we can derive the eigenvalue for the Hamiltonian, eigenfunctions of L, such as $\varphi_\alpha(p)\varphi_\alpha(p)$, correspond to zero eigenvalues of the Liouville operator, $E_\alpha - E_\alpha = 0$, and are therefore invariants of motion. As a result the system is integrable (as is a system of noninteracting particles), and cannot reach equilibrium. This is a form of the quantum paradox.

We can now see clearly why it is not sufficient to extend wave functions beyond Hilbert space. Indeed, as indicated in Section III, this leads to complex energies in the form $E_\alpha = \omega_\alpha - i\gamma_{\alpha,}$ where ω_α is the real past and γ_α the life span, which describe the decay of excited atoms or unstable particles, but this still does not account for irreversible processes associated with the approach to equilibrium. In spite of the complex element in E_α, all diagonal elements of ρ, which are products such as $\varphi_\alpha(p)\varphi_\alpha(p')$, would be invariants because the eigenvalue $E_\alpha - E_\alpha$ again vanishes, and the system remains integrable and cannot approach equilibrium.*

The experimental basis of Bohr's theory of atoms and the subsequent emergence of quantum theory is based on the Ritz-Rydberg principle, according to which each frequency ν, as measured in spectroscopy, is the difference between the two numbers E_α and E_β, which represent two quantum levels. This, however, can no longer be true for systems presenting irreversible processes that lead the system to equilibrium. Quantum theory must therefore be fundamentally revised.

Historically, the roots of mechanics lie in two branches of physics: the thermal equilibrium between matter and radiation that led Planck to introduce his famous constant h in 1900, and spectroscopy, which led from the Ritz-Rydberg principle to Bohr's atom, and finally, with Heisenberg (1926), to quantum theory. However, the relationship between these two domains has never been elucidated. We see that the Ritz-Rydberg principle is incompatible with the thermal approach to equilibrium described by Planck's work. Thus we need a new formulation making thermal physics and spectroscopy compatible. This can be

---

*Difficulties arise when $E_\alpha - E_\beta$ is replaced by $E_\alpha - E^*_\beta$, where E^*_β is the complex conjugate of E_β. Here, $E_\alpha - E^*_\alpha = -2i\gamma_\alpha \neq 0$, with no equilibrium state.

achieved at the level of probability distributions from which we may derive observable frequencies (including their complex part), but these frequencies are no longer differences in energy levels for the systems we expect to approach equilibrium. We have to solve the quantum Liouville eigenvalue problem for LPS in the context of more general function spaces. As in classical mechanics, this will involve two basic ingredients: delocalized distribution functions, which lead to singularities, and Poincaré resonances, which lead to new dynamical processes. As in classical dynamics, there then appear new solutions at the statistical level that cannot be reduced to the traditional wave function formalism of quantum mechanics, and no longer satisfy the Ritz-Rydberg principle. In this sense, we can truly speak of a new formulation of quantum theory.

V

With certain modifications, we can follow the probabilistic formulation for classical systems given in Chapter 5. The formal solution of the Liouville equation is $i(\partial\rho/\partial t) = L\rho$, where in quantum theory $L\rho$ is the commutator of the Hamiltonian with ρ (as we have seen, $L\rho = H\rho - \rho H$). It can be written as either $\rho(t) = e^{-iHt}\rho(0)e^{+iHt}$ or $\rho(t) = e^{-iLt}\rho(0)$. What is the difference between these equations? In the first formulation, it appears that we would have two *independent* dynamic evolutions: one associated with e^{-iHt} and the other with e^{+iHt}, one moving toward the "future" and the other toward the "past" (as t is replaced by $-t$). If this were so, we could expect no time-symmetry breaking, and the statistical description would conserve the time symmetry of the Schrödinger equation. But this is no longer the case when we include Poincaré resonances, which couple the two time evolutions (e^{-iHt} and e^{+iHt}). There is now only one independent time evolution (time has "one dimension"). In order to study time-symmetry breaking, we have to begin with the expression $\rho(t) = e^{-itL}\rho(0)$, which describes a single time sequence in the Liouville space. In other words, we have to order dynamical events according to a single time sequence.* We can then describe interactions, as we did for classical mechanics, as a succession of events separated by free motion. In classical mechanics, these events change the values of the wave vector k and the momenta p. In Chapter 5, we introduced various events leading to the creation and destruction of correlations, where we saw that the decisive factor was the appearance, for LPS, of new events (the bubbles in Figure 5.7) that couple creation and destruction. As such, they radically change classical dynamics because they introduce diffusion, break determinism, and destroy time symmetry. We can also identify the same events in quantum mechanics. To do so, we need to introduce variables that play the same role as the wave vector k in classical theory's Fourier representation. In classical mechanics, we start with a statistical formulation in which the distribution functions $\rho(q, p)$ are expressed as functions of the coordinates q and the momenta p. We then proceed to the Fourier transformation $\rho_k(p)$ involving the wave vector k and the momenta.

In quantum mechanics, we can follow a similar procedure.[10] We start with the density matrix $\rho(p, p')$ in the mo-

---

*If this is not done, we have to be very careful. Feynman's well-known statement that an electron propagates toward the future and a position moves toward the past refers to time as it appears in the Schrödinger equation before ordering dynamical events according to a single time sequence.

mentum representation, which is a function of two sets of variables, p and p'. We then introduce new variables, $k = p - p'$ and $P = (p + p')/2$; we can now write, as in classical mechanics, $\rho_k(P)$. It can then be shown that k plays the same role in quantum mechanics as the wave vector does in classical mechanics. (For example, in interactions, the sum of the wave vectors is conserved, that is, $k_j + k_n = k'_j + k'_n$.) Again as in classical mechanics, Poincaré resonances introduce new dynamical events that couple the creation and destruction of correlations, and therefore describe quantum diffusive processes.

The formulation of classical and quantum theory for LPS is more or less parallel. A minor difference appears in the role of the momentum P. For each event, as introduced in Chapter 5, the momenta of the interacting particles are altered. In quantum mechanics, we use the two variables k and P, where the variable P replaces the classical momentum. As these variables interact, the modification of P involves Planck's constant h. For $h \to 0$, however, we come back to the classical momentum p. But this difference has no important effect on formal development, and we shall not attempt to describe it in further detail.

In the previous chapter, we introduced a fundamental difference between transitory and persistent interactions. Persistent interactions are especially significant because they appear in all situations where thermodynamics can be applied. As in classical mechanics, the distribution function ρ corresponding to persistent interactions is described by singular functions of the variable k. In classical dynamics, as well as classical and quantum mechanics, persistent scattering is typical of the situations described by statistical mechanics and cosmology. For example, in the atmosphere, particles collide continuously, are scattered, and then recollide. Persistent scattering is described by delocalized distribution functions, which are singular functions in the wave vector space. As we saw in Chapter 5, the latter force us to go outside Hilbert space.

By taking into account delocalized singular distribution functions and Poincaré resonances, we obtain, as in classical mechanics, complex, irreducible spectral representations for the Liouville operator L. Again, as in classical dynamics, irreversibility is associated with the appearance of higher- and higher-order correlations. As in classical mechanics, this leads to new features in kinetic theory and macroscopic physics. The basic conclusions of our formulation of quantum mechanics are as follows:

• The eigenvalues of the Liouville operator are no longer differences between the eigenvalues of the Hamiltonian, which are obtained from the Schrödinger equation. Therefore, the Ritz-Rydberg principle is violated, whereby the systems are no longer integrable and the approach to equilibrium is possible.
• The quantum superposition principle associated with the linearity of the Schrödinger equation is violated.
• The eigenfunctions of the Liouville operator are not expressed in terms of probability amplitudes or wave functions, but rather in terms of probabilities proper.

Our predictions have already been verified in simple situations where we can follow the collapse of wave functions outside Hilbert space.[11] Moreover, they have led to interesting predictions of the form of spectral lines, and have allowed us to accurately describe the approach to equilibrium. We regret that we cannot go into greater detail about their specific applications, but our objective in this book is merely to provide a brief tour of the theoretical background.

VI

At the Fifth Solvay Conference on Physics that took place in Brussels in 1927, there was an historic debate between Einstein and Bohr. In the words of Bohr:

> To introduce the discussion on such points, I was asked at the conference to give a report on the epistemological problems confronting us in quantum physics and took the opportunity to center upon the question of an appropriate terminology and to stress the viewpoint of complementarity. The main argument was that unambiguous communication of physical evidence demands that the experimental arrangement as well as the recording of the observations be expressed in common language, suitably refined by the vocabulary of classical physics.[12]

But how can we describe an apparatus in classical terms in a world dominated by quantum laws? This is the weak point in the so-called Copenhagen interpretation. Nevertheless, there is an important element of truth contained therein. Measurement is a means of communication. It is because we are both "actors and spectators," to use Bohr's words, that we can learn something about nature. But communication requires a common time. The existence of this common time is one of the basic consequences of our approach.

The apparatus that performs the measurements, whether a physical construct or our own sensory perception, must follow the extended laws of dynamics, including time-symmetry breaking. There do exist integrable time-reversible systems, but we cannot observe them in isolation. As emphasized by Bohr, we need an apparatus that breaks time symmetry. LPS blur this distinction in that they already break time symmetry and therefore, in a sense, measure themselves. We do not have to describe an apparatus in classical terms. Common time emerges at the quantum level for LPS associated with thermodynamic systems.

The subjective aspect of quantum theory, which attributed an unreasonable role to the observer, deeply troubled Einstein. To our way of thinking, through his measurements the observer no longer plays some extravagant role in the evolution of nature—at least no more so than in classical physics. We all transform information received from the outside world into actions on a human scale, but we are far from being the demiurge, as postulated by quantum physics, who would be responsible for the transition from nature's potentiality to actuality.

In this sense, our approach restores sanity. It eliminates the anthropocentric features implicit in the traditional formulation of quantum theory. Perhaps this would have made quantum theory more acceptable to Einstein.

Chapter 7

OUR DIALOGUE
WITH NATURE

I

Science is a dialogue between mankind and nature, the results of which have been unpredictable. At the beginning of the twentieth century, who would have dreamed of unstable particles, an expanding universe, self-organization, and dissipative structures? But what makes this dialogue possible? A time-reversible world would also be an unknowable world. Knowledge presupposes that the world affects us and our instruments, that there is an interaction between the knower and the known, and that this interaction creates a difference between past and future. Becoming is the *sine qua non* of science, and indeed, of knowledge itself.

The attempt to understand nature remains one of the

basic objectives of Western thought. It should not, however, be identified with the idea of control. The master
who believes he understands his slaves because they obey
his orders would be blind. When we turn to physics, our
expectations are obviously very different, but here as well,
Vladimir Nabokov's conviction rings true: "What can be
controlled is never completely real; what is real can never
be completely controlled."[1] The classical ideal of science, a
world without time, memory, and history, recalls the total-
itarian nightmares described by Aldous Huxley, Milan
Kundera, and George Orwell.

In our recent book, *Entre le Temps et l'Eternité,* Isabelle
Stengers and I wrote:

> Perhaps we need to start by emphasizing the almost incon-
> ceivable character of dynamic reversibility. The question of
> time—of what its flow preserves, creates and destroys—has
> always been at the center of human concerns. Much specula-
> tion has called the idea of novelty into question and affirmed
> the inexorable linkage between cause and effect. Many forms
> of mystical teaching have denied the reality of this changing
> and uncertain world, and defined an ideal existence permit-
> ting escape from life's afflictions. We know how important
> the idea of cyclical time was in antiquity. But, like the rhythm
> of the seasons or the generations of man, this eternal return
> to the point of origin is itself marked by the arrow of time.
> No speculation, no teaching has ever affirmed an equivalence
> between what is done and what is undone: between a plant
> that sprouts, flowers and dies, and a plant that resuscitates,
> grows younger and returns to its original seed; between a
> man who grows older and learns, and one who becomes a
> child, then an embryo, then a cell.[2]

In Chapter 1, we alluded to Epicurus' dilemma and the atomistic approach of the ancients. Today, the situation has changed significantly in the sense that the more we know about our universe, the more difficult it becomes to believe in determinism. We live in an evolutionary universe whose roots, which lie in the fundamental laws of physics, we are now able to identify through the concept of instability associated with deterministic chaos and nonintegrability. Chance, or probability, is no longer a convenient way of accepting ignorance, but rather part of a new, extended rationality. As we have seen, for these systems, the equivalence is broken between the individual description (trajectories and wave functions) and the statistical description (in terms of ensembles). At the statistical level, we can incorporate instability. The laws of nature, which no longer deal with certitudes but possibilities, overrule the age-old dichotomy between being and becoming. They describe a world of irregular, chaotic motions more akin to the image of the ancient atomists than to the world of regular Newtonian orbits. This disorder constitutes the very foundation of the macroscopic systems to which we apply an evolutionary description associated with the second law, the law of increasing entropy.

We have considered deterministic chaos, and we have discussed the role of Poincaré resonances in both classical and quantum mechanics. We have seen that we need two conditions to obtain our statistical formulations, which go beyond the usual ones for classical and quantum mechanics: the existence of Poincaré resonances, which lead to new diffusion–type processes that can be incorporated into the statistical description, and extended persistent interactions described by delocalized distribution functions.

These conditions lead to a more general definition of chaos. As in the case of deterministic chaos, we then obtain new solutions for the statistical equations that cannot be expressed in terms of trajectories or wave functions. If these conditions are not satisfied, we return to the usual formulations. This is the case in many simple examples, such as two-body motion (for instance, the sun and earth), and typical scattering experiments, where before and after scattering, the particles are free. These examples, however, correspond to idealizations. The sun and earth are part of the many-body planetary system; scattered particles will eventually meet other particles, and are therefore never free.

It is only by isolating a certain number of particles and studying their dynamics that we obtain the traditional formulations. Conversely, time-symmetry breaking is a global property encompassing Hamiltonian dynamical systems as a whole. In the chaotic maps studied in Chapters 3 and 4, irreversibility occurs even in systems with few degrees of freedom due to the simplifications used to describe the equations of motion.

A remarkable feature of our approach is its application to both classical and quantum systems. All other theoretical proposals that we are aware of attempt to eliminate the quantum paradox through an exclusively quantum mechanism. On the contrary, in our view, the quantum paradox is only one aspect of the time paradox. In the Copenhagen interpretation, the need to introduce two different types of time evolution is engendered by the measurement process. According to Bohr himself, "Every atomic phenomenon is closed in the sense that its observation is based on a recording obtained by means of suitable amplification devices with irreversible functions, such as permanent marks on a photographic plate."[3] It was this measurement problem that led to the need for a collapse of the wave function, and forced us to introduce a second type of dynamical evolution into quantum mechanics. It is therefore not surprising that the time paradox and quantum paradox are so closely linked. In solving the former, we also solve the latter. As we have seen for LPS, quantum dynamics can only be described at the statistical level. Moreover, to learn something about quantum processes, we again need an LPS acting as an apparatus. It is thus the second law of quantum time evolution, which includes irreversibility, that becomes the general one.

As stated by Alastair Rae, "A pure quantum process (described by the Schrödinger equation) occurs only in one or more parameters that have become detached from the rest of the universe, and perhaps even from space-time itself, and leave no trace of their behavior on the rest of the universe until a measurement interaction takes place."[4] Whatever the process, at some point irreversibility has to come into the picture. An almost identical statement could have been made regarding classical mechanics!

It has often been said that in order to make progress in these difficult areas, we need the inspiration of a truly crazy idea. Heisenberg was fond of asking what the difference is between an abstract painter and a good theoretical physicist. In his opinion, an abstract painter needs to be just as original as a good theoretical physicist needs to be conservative.[5] We have tried to follow Heisenberg's advice. Our line of reasoning in this book is certainly less radical than most other attempts made in the past to solve the time or quantum paradox. Perhaps our craziest idea is that trajectories are not primary objects, but rather the result of a superposition of plane waves. Poincaré resonances destroy the coherence of these superpositions, and lead to an irreducible statistical description. Once this is understood, the generalization to quantum mechanisms becomes easy.

II

Numerous references have been made in this text to the thermodynamic limit, which is defined by the limit N (number of particles) $\rightarrow \infty$, and volume $V \rightarrow \infty$, where the concentration N/V remains finite. This limit simply means that when the number of particles N is sufficiently large, terms such as $1/N$ can be ignored. This is true for the usual thermodynamic systems where N is typically on the order of 10^{23}. However, there are no systems that contain an infinite number of particles.

The universe itself is highly heterogeneous and far from equilibrium. This prevents systems from reaching a state of equilibrium. For example, the flow of energy that originates in the irreversible nuclear reactions within the sun maintains our ecosystem far from equilibrium, and has thus made it possible for life to develop on earth. As we saw in Chapter 2, nonequilibrium leads to new collective effects and to a new coherence. It is interesting that these are precisely the consequences of the dynamical theory presented in Chapters 5 and 6.

There are two types of effects produced by nonequilibrium. If, as in the Bénard instability, we heat a liquid from below, we produce collective flows of molecules. When we stop the heating process, the flows disintegrate and return to the usual thermal motion. In chemistry, the situation is different; irreversibility leads to the formation of molecules that cannot be produced in near-equilibrium conditions. In this sense, irreversibility is inscribed in matter. This is likely to be the origin of self-replicating bio-molecules. While we shall not pursue this question here, let us merely note that molecules of comparable complexity can indeed be produced, at least through computer simulations, in nonequilibrium conditions.[6] In the next chapter, which discusses cosmology, we argue that matter itself is the result of irreversible processes.

In nonrelativistic physics, whether classical or quantum, time is universal, but the flow of time as associated with irreversible processes is not. It is to the fascinating implications of this distinction that we shall now turn.

III

Let us first consider a chemical model. If we start at time t_0 with two identical samples of mixtures of two gases, such as carbon monoxide (CO) and oxygen (O_2), a chemical reaction leading to carbon dioxide (CO_2) can be catalyzed by metallic surfaces. In one of the samples, we introduce such a catalyst, and in the other, we do not. If we compare the two samples at a later time t, their composition will therefore be quite different. The entropy produced in the sample containing the catalytic surface will be much greater as a result of the chemical reaction. If we associate the production of entropy to the flow of time, time itself will appear to vary between the two samples. This observation is in agreement with our dynamical description. The flow of time is rooted in Poincaré resonances that depend on the Hamiltonian, that is, on dynamics. The introduction of a catalyst changes the dynamics, and therefore alters the microscopic description. In another example, gravitation again changes the Hamiltonian, and therefore the resonances. We then have a kind of nonrelativistic analogue of the twin paradox of relativity, which we shall come back to in Chapter 8. For the moment, suppose that we send two twins (who are simply two LPS) into space, leaving the earth at t_0 and coming back at t_1 (see Figure 7.1). Before their return, one twin goes through a gravitational field, and the other does not. The entropy produced (as a result of Poincaré resonances) will be different, and our twins will come back with different "ages," leading us to the basic conclusion that the flow of time, even in a Newtonian universe, may have different effects according to the processes considered. Our conclusion is in stark contrast with the Newtonian view, which was based on a universal flow of time. But what can a flow of time mean in a description of nature in which past and future play the same role? It is irreversibility that leads to a flow of time. Time evolution is no longer described by *groups* where past and

Figure 7.1
Effect of a Gravitational Field on the Flow of Time

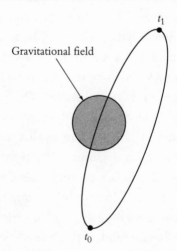

future play the same role, but rather by *semigroups* that include the direction of time. When we introduce a time associated with the production of entropy (see Chapter 2), as the sign of entropy production is positive, entropic time always points in the same direction. This is the case in the two examples mentioned above even though entropic time does not keep pace with clock time.

We could introduce an "average" entropic time for the entire universe, but this would not have a great deal of meaning because of the heterogeneity of nature. Irreversible geological processes have a time scale distinct from those of biological processes. Even more important, there exists a multiplicity of evolutions, which are particularly evident in the field of biology. As stated by Stephen J. Gould, bacteria have remained basically the same since the Precambrian era, while other species have evolved dramatically, often over short time scales.[7] It would therefore be a mistake to consider a simple one-dimensional evolution. Some two hundred million years ago, certain reptiles started to fly, while others remained on earth. At a later stage, certain mammals returned to the sea, while others remained on land. Similarly, certain apes evolved into humanoids, while others did not.

At the conclusion of this chapter, it is appropriate to cite Gould's definition of the historical character of life:

> To understand the events and generalities of life's pathway, we must go beyond principles of evolutionary theory to a paleontological examination of the contingent pattern of life's history on our planet—the single actualized version among millions of plausible alternatives that happened not to occur. Such a view of life's history is highly contrary both to con-

ventional deterministic models of Western science and to the
deepest social traditions and psychological hopes of Western
cultures for a history culminating in humans as life's highest
expression and intended planetary steward.[8]

We are in a world of multiple fluctuations, some of
which have evolved, while others have regressed. This is in
complete accord with the results of far-from-equilibrium
thermodynamics obtained in Chapter 2. But we can now
go even farther. These fluctuations are the macroscopic
manifestations of fundamental properties of fluctuations
arising on the microscopic level of unstable dynamical
systems. The difficulties emphasized by Gould are no
longer present in our statistical formulation of the laws of
nature. Irreversibility, and therefore the flow of time, starts
at the dynamical level. It is amplified at the macroscopic
level, then at the level of life, and finally at the level of
human activity. What drove these transitions from one
level to the next remains largely unknown, but at least we
have achieved a noncontradictory description of nature
rooted in dynamical instability. The descriptions of nature
as presented by biology and physics now begin to con-
verge.

Why does a common future exist at all? Why is the
arrow of time always pointed in the same direction? This
can only mean that our universe forms a whole. It has
a common origin that already implied time-symmetry
breaking. Here we encounter cosmological problems. In
dealing with them, we must embrace gravity and enter the
world of Einstein's theory of relativity.

Chapter 8

DOES TIME PRECEDE
EXISTENCE?

I

Several years ago, I delivered a physics colloquium at Lomonosoff University in Moscow. Afterwards, Professor Ivanenko, one of the most respected Russian physicists, asked me to write a short inscription on a particular wall where there were already many sentiments expressed by famous scientists such as Dirac and Bohr. I vaguely remember the sentence chosen by Dirac, which was something like: "Beauty and truth go together in theoretical physics." After some hesitation, I wrote: "Time precedes existence."

For many physicists, the acceptance of the big bang theory as the origin of our universe means that time must have a beginning, and perhaps an end. It seems more likely to me that the birth of our universe was only one event in the history of the entire cosmos, and that we therefore have to ascribe to that so-called "meta-universe" a time prior to the birth of our own.

We know that we are living in an expanding universe. The *standard model,* which dominates the field of cosmology today, asserts that if we were to go backward in time, we would arrive at a singularity, a point that contains the totality of the energy and matter in the universe. However, the model does not enable us to describe this singularity because the laws of physics cannot be applied to a point corresponding to an infinite density of matter and energy. It is no wonder that John Archibald Wheeler speaks of the big bang as confronting us "with the greatest crisis in physics."[1] Can we accept the big bang as a real event, and how is it possible to reconcile this event with laws of nature that are time reversible and deterministic? We come back to the problems of measurement and irreversibility, but now in the cosmological context.

Since the discovery of the big bang, the scientific community has reacted to the strange nature of this singularity by attempting to eliminate the big bang entirely (see the steady-state theory in Sections I and III), or considering it as a kind of "illusion" arising from the use of an incorrect concept of time (see Hawking's imaginary time in Section II), or even viewing it as a sort of miracle akin to the biblical description in Genesis.

As we have already noted, it is impossible to discuss cosmology today without referring to the theory of relativity, "the most beautiful theory in physics," according to the celebrated textbook by Lev Davidovich Landau and Evgeny Mikhailovich Lifschitz.[2] In Newtonian physics, even when extended by quantum theory, space and time are given once and for all. Moreover, there is a universal time common to all observers. In relativity, this is no longer the case; space and time are now part of the picture. What consequences does this have for our own interpretation? In his recent book, *About Time,* Paul C. W. Davies comments on the impact of relativity, "The very division of time into past, present and future seems to be physically meaningless."[3] He repeats Hermann Minkowski's famous statement: "Henceforth space by itself, and time by itself, are doomed to fade away into mere shadows."[4]

We have already alluded to Einstein's celebrated assertion that "for us convinced physicists, the distinction between past, present and future is an illusion, although a persistent one."[5] At the end of his life, however, Einstein seems to have changed his mind. In 1949, he was offered a collection of essays that included a contribution by the great mathematician Kurt Gödel, who had taken quite seriously Einstein's statement that time as irreversibility was only an illusion. When he presented Einstein with a cosmological model in which it was possible to return to one's own past, Einstein was not enthusiastic. In his answer to Gödel, he wrote that he could not believe that he could "telegraph back to his own past." He even added that this impossibility should lead physicists to reconsider the problem of irreversibility.[6] That is precisely what we have attempted to do.

In any case, we wish to emphasize that the revolution brought about by relativity in no way affects our previous conclusions. Irreversibility, or the flow of time, remains as "real" as in nonrelativistic physics. Perhaps we could argue that irreversibility plays an even greater role when we go to higher and higher energies. It has been suggested, mainly by Hawking, that in the early universe, space and time lose their distinction, and time becomes fully "spatialized." But no one to our knowledge has devised a mechanism for this spatialization of time, or a means by which space and time could emerge from what is often described as a "foamy mess."

Our position is quite different from those stated above in that we consider the big bang an irreversible process *par excellence*. We suggest that there would have been an irreversible phase transition from a preuniverse that we call the *quantum vacuum*. This irreversibility would result from an instability in the preuniverse induced by the interactions of gravitation and matter. Clearly we are at the edge of positive knowledge, even dangerously close to science fiction.

Nevertheless, we argue that irreversible processes associated with dynamical processes have probably played a decisive role in the birth of our universe. From our perspective, time is eternal. We have an age, our civilization has an age, our universe has an age, but time itself has neither a beginning nor an end. This brings closer two of the traditional views of cosmology: the steady-state theory introduced by Hermann Bondi, Thomas Gold, and Fred Hoyle, which may apply more precisely to the unstable medium that generates our universe (the meta- or preuniverse), and the standard big bang approach.[7]

Again, speculative elements cannot be avoided, but we find it interesting that views emphasizing the role of time and irreversibility can be formulated more precisely than before, even though the ultimate truth is still far beyond our reach. We agree entirely with the Indian cosmologist Jayant Vishnu Narlikar, who wrote, "Astrophysicists of today who hold the view that the 'ultimate cosmological problem' has been more or less solved may well be in for a few surprises before this century is out."[8]

II

As we proceed with our investigation, let us consider Einstein's *special relativity.* This theory takes as its starting point two inertial observers moving at a constant velocity with respect to one another. In prerelativistic, Galilean physics, it was accepted that the distance between the two observers, $l_{12}^2 = (x_2-x_1)^2 + (y_2-y_1)^2 + (z_2-z_1)^2$, would remain the same as the difference between the two instants, $(t_2 - t_1)^2$. Spatial distance was defined in terms of Euclidean geometry. This, however, led to different values of the velocity of light c in the vacuum as measured by the two observers. In accordance with our experience, if we assume that both observers measure the same value of the velocity of light, we must introduce (as did Lorentz, Poincaré, and Einstein) the spatiotemporal interval, $s_{12}^2 = c^2(t_1-t_2)^2-l_{12}^2$. It is this interval that is conserved when we move from one inertial observer to the other. In contrast to Euclidean geometry, we now have the Minkowski space-time interval. The transition from one coordinate system, $x, y, z, t,$ to another, $x', y', z', t',$ is the famous Lorentz transformation that combines space and time. At no point, however, is the distinction between space and time lost; in the spatiotemporal interval, the minus sign indicates space dimensions, and the plus sign indicates time.

This situation is often illustrated by the spatiotemporal diagram represented in Figure 8.1. On one axis there is time $t,$ and on the other a single geometrical coordinate $x.$ In relativity, the velocity of light c in the vacuum is the maximum speed at which signals can be transmitted. We can therefore distinguish among different regions in the diagram.

In this diagram the observer is situated at O. His future is included in the "cone" *BOA,* and his past in the cone *A'O'B'.* These cones are determined by the velocity of light *c* in that the velocities inside them are smaller than *c,* and outside them are greater, and therefore impossible to realize. In this diagram, the event *C* is simultaneous with O, while event *D* precedes O. But this conclusion is purely conventional because a Lorentz transformation would rotate the axis *t, x,* in which case *D* might appear as simultaneous with O, and *C* posterior to O. Simultaneity is modified by the Lorentz transformation, but the cone of light is *not.* The direction of time is thus invariant. The problem of ascertaining whether or not the laws of nature are time symmetric remains essentially the same in relativity as in prerelativistic physics, but now this question is even more pertinent. At best, O knows all the events that occurred in his past, that is, in the cone *A'O'B'.* As represented in Figure 8.2, events starting in *C* or *D* will reach him only at later times, t_1 and t_2, even if they are associated

Figure 8.1

Distinction Between Future and Past in Special Relativity

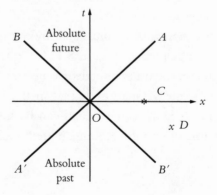

Figure 8.2

Events starting at C and D will reach the observer O at future times t_1 and t_2.

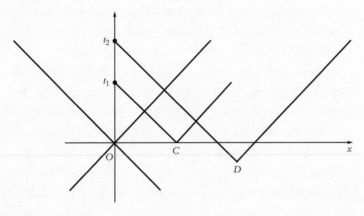

with signals traveling at the velocity of light. As a result, O can collect only limited data. In an amusing analogy with deterministic chaos made by Baidyanath Misra and Ioannis Antoniou, it is said that a relativistic observer has only a finite window on the outside world, and here also a deterministic description corresponds to an overidealization.[9] This gives us yet another reason to proceed to a statistical description.

There are, of course, most interesting new effects introduced by relativity, such as the famous twin paradox, where one twin remains on earth at point $x = 0$, while the other leaves in a spaceship that changes direction at t_0 (in the coordinate system in which O is at rest), and comes back to earth at $2t_0$. The time interval, as measured by the moving twin, is greater than $2t_0$. This is Einstein's remarkable time dilation prediction, which has been verified by using unstable particles. The lifetime of these twins therefore depends on the path as predicted by relativity. In Chapter 7, we stated that the flow of time depends on a history of events, but Newtonian time is universal and independent of history. Now time itself becomes history dependent.

In his seminal book, *The Theory of Space, Time and Gravitation,* Vladimir A. Fock emphasizes that we have to be extremely careful when discussing the twin paradox inasmuch as the effect of acceleration on the clock in the moving spaceship is neglected.[10] He shows that when we consider a more detailed model in which acceleration is due to a gravitational field described by general relativity, different results are obtained. The sign of time dilation can even be changed. These predictions of general relativity should lead to fascinating new experiments to test their validity.

In his *Brief History of Time,* Hawking introduces imaginary time, $\tau = it$, where all four dimensions are "spatialized" in the Minkowski spatiotemporal interval.[11] According to Hawking, real time may well be this imaginary time, whereby the mathematical formula for the Lorentz interval becomes symmetric. Hawking's proposition does indeed go beyond relativity, but it is only one more attempt to negate the reality of time in describing the universe as a static, geometrical structure, in contradiction to the role that the flow of time plays at all levels of observation.

Let us now come back to the crux of our argument and consider the effect of relativity on the systems described by classical Hamiltonian dynamics or quantum mechanics. Dirac, and others who came after him, showed how to combine the requirements of special relativity with a Hamiltonian description.[12] Relativity dictates that the laws

Figure 8.3

The Twin Paradox

Observer O' is in motion in relation to observer O.

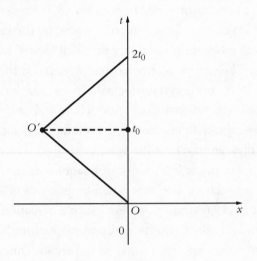

of physics remain the same for all inertial systems. In Chapters 5 and 6, we assumed implicitly that the systems as a whole are at rest. But according to relativity, a similar description is valid whether or not the system as a whole is moving at uniform velocity with respect to some observer. We have seen that Poincaré resonances destroy the dynamical group in which past and future play the same role, whereby we obtain semigroups that break time symmetry. In prerelativistic physics, the groups and semigroups maintain the distance l_{12}^2 invariant. In relativistic theory, we can introduce as well both groups and semigroups which leave invariant the Minkowski interval. Unfortunately, the proof is too technical to be given here. In any case, this conclusion shows that the Minkowski space-time interval is no way in contradiction to irreversible processes. It is not true that relativity implies the spatialization of time. As stated by Minkowski, space and time are no longer independent entities, but this does not preclude the existence of an arrow of time.

Such a conclusion could be anticipated. If time-symmetry breaking occurs in one inertial frame, by the very definition of relativity, it has to appear in all inertial reference systems. The theory of irreversible processes is thus quite similar (apart from certain formal changes) in both nonrelativistic and relativistic systems. There is, however, one basic difference: Interactions are no longer instantaneous; rather, they propagate at the velocity of light. For charged particles within the framework of quantum theory, for example, interactions are transmitted by photons. This leads to additional irreversible processes such as radiation damping, which results from the emission of photons by particles. In more general terms, in relativistic physics we consider particles as associated with fields (the photons are the particles associated with the electromagnetic field), and irreversibility results from the interaction of these fields.

Until now we have considered the Minkowski spacetime interval as it corresponds to special relativity. In order to complete our discussion of cosmology, we have to include gravitation, which first requires a generalization of the space-time interval.

III

Let us first return to the question of the big bang. As we mentioned above, by following our expanding universe backward in time, we come to a singularity in which density, temperature, and curvature all become infinite. From the rate of recession of the galaxies as observed today, we can estimate that the birth of the universe occurred approximately fifteen billion years ago. This period of time that separates us from the big bang is surprisingly short. To express it in years, we use the rotation of the earth as a clock. Fifteen billion revolutions is indeed a small number if we remember that in the hydrogen atom, the electron rotates some 10,000 billion times per second!

Whatever the time scale, the existence of a primordial event at the origin of our universe is certainly one of the most extraordinary suggestions science has ever made. Physics deals only with classes of phenomena, and the big bang does not seem to belong to any of these. At first view, it appears to have no parallel elsewhere in physics.

Many scientists have been willing to explain this singularity in terms of the "hand of God," or the triumph of the biblical story of creation, whereby science would reconstruct the existence of an act that transcends physical rationality. Others have tried to avoid what they see as a disquieting situation. One remarkable attempt in this sense is the steady-state universe proposed by Bondi, Gold, and Hoyle.[13] This model is based on the perfect cosmological principle: Not only is there no privileged place in the universe, but there is also no privileged time. According to this principle, every observer, in the past and in the future, is able to attribute to the universe the same values of parameters such as temperature and matter density. The steady-state universe is characterized by an exponential expansion compensated by a permanent creation of matter. The synchronization between expansion and creation maintains a constant density of matter-energy, and thus leads to the image of an eternal universe in a state of continous creation. In spite of its appeal, the steady-state model implies certain major difficulties. In particular, in order to maintain the steady state, we need a fine-tuning between cosmological evolution (the expansion of the universe) and microscopic events (the creation of matter). As long as no mechanism for this is proposed, the hypothesis of compensation between expansion and creation is highly questionable.

It was an experimental result that led the great majority of cosmologists to reject the steady-state model in favor of the big bang, which is now considered the standard model. This occurred in 1965, when Arno Penzias and Robert Wilson identified the now-famous fossil radiation at 2.7 K.[14] The existence of such radiation had been predicted as early as 1948 by Ralph A. Alpher and Robert Herman, who reasoned that if the universe was much hotter and denser in the past than it is today, then it must have been "opaque," with photons possessing sufficient energy to interact strongly with matter. It can be shown that at a temperature of approximately 3,000 K, the equilibrium between matter and light is destroyed, and our universe becomes transparent as radiation is "detached" from matter. The only subsequent change in the properties of the photons that form the thermal radiation is the change in their wavelength, which increases with the size of the universe. Alpher and Herman were thus able to predict that if the photons indeed formed a black body radiation at 3,000 K at the time when their equilibrium with matter was destroyed (that is, some 300,000 years after the "origin"), the temperature of this radiation should correspond today to a temperature of about 3 K. This was a landmark prediction that anticipated one of the greatest experimental findings of this century.[15]

The standard model is very much at the core of present-day cosmology, and scientists generally accept that it leads to a correct description of the universe starting one second after the big bang singularity. But the state of the universe during its first second of life still remains an open question.

Why is there something rather than nothing? This appears to be the ultimate question beyond the range of positive knowledge. However, this question can be formulated in physical terms, and thereby linked to the problem of instability and time. One such formulation that has become quite popular today defines the birth of our universe as a *free lunch*. Edward Tryon presented this idea in 1973, but it seems to hark back to Pascual Jordan. In Tryon's view, our universe can be described as having two forms of energy: one related to attractive gravitational forces, which is negative, and the other related to mass according to Einstein's celebrated formula $E = mc^2$, which is positive.[16]

It is tempting to speculate that the total energy of the universe could be zero, as is the energy of an empty universe. The big bang would thus be associated with fluctuations in the vacuum conserving the energy. This is truly an appealing idea. The generation of nonequilibrium structures (such as Bénard vortices or chemical oscillations), where energy is conserved, also corresponds to a free lunch, for the price of nonequilibrium structures is entropy, and not energy. In this context, can we specify the origin of negative gravitational energy and its transformation into positive matter-energy? This is the question that we shall now address.

IV

Perhaps Einstein's most profound contribution was to associate gravitation with the curvature of space-time. As we have seen in special relativity, the Minkowski space-time interval is $ds^2 = c^2dt^2 - dl^2$. In general relativity, the space-time interval becomes $ds^2 = \Sigma g_{\mu\nu}dx^\mu dx^\nu$, where μ, ν take on four values: 0 (time), and 1, 2, 3 (space). The ten distinct functions obtained (given that $g_{\mu\nu} = g_{\nu\mu}$) characterize space-time, or Riemannian geometry. A simple example that illustrates Riemannian geometry is a sphere considered as a curved two-dimensional space.

In the Newtonian view, space-time is given once and for all, independent of the matter it contains. Now we understand, thanks to the Einsteinian revolution, that the connection between space-time and matter is expressed by Einstein's fundamental field equations, which relate two objects: On the one hand, we have an expression that describes the curvature of space-time in terms of the $g_{\mu\nu}$ and its derivatives with respect to space and time, and on the other an expression that defines the material content in terms of its matter-energy content and pressure. This material content is the source of the curvature of space-time. Einstein applied his equations to the universe as a whole as early as 1917, and in so doing, set the course of modern cosmology. To achieve this application, he developed a timeless static model in accord with his philosophical views. Baruch Spinoza was Einstein's favorite philosopher, and we can recognize his spirit in the choice of the model.

Then came a succession of surprises. Alexander Friedmann and Georges-Henri Lemaître proved that Einstein's universe was so unstable, the smallest fluctuation would destroy it.[17] On the experimental side, Edwin Powell Hubble and his colleagues discovered the expansion of our universe.[18] Then in 1965 came the observation of residual black body radiation, which led to the present standard cosmological model.

In order to go from the basic equations of general relativity to the field of cosmology, we have to introduce simplifying assumptions. The standard model associated with Alexander Friedmann, Georges-Henri Lemaître, Howard Robertson, and Arthur Walker is founded on the cosmological principle that the universe, when viewed on a large scale, may be considered homogeneous and isotropic. The metrics thus take on the far simpler form $ds^2 = c^2dt^2 - R(t)^2dl^2$ (the so-called Friedmann interval). This expression differs from Minkowski space-time in two respects: dl^2 is a spatial element that corresponds to either a zero-space curvature (as in the Minkowski space) or to a positive or negative curvature (as in a sphere or hyperboloid). $R(t)$, which is usually called the radius of the universe, corresponds to the limit of astronomical observations at time t. Einstein's equations relate $R(t)$ and the space curvature to the average density and pressure of the energy-matter. Einstein's cosmological evolution is also formulated as conserving entropy, and his equations are consequently time reversible.

It is generally accepted that the standard model permits us to understand at least qualitatively what happened to our universe a fraction of a second after its birth. This is an extraordinary achievement, but we are still left with the question of what occurred before. When we extrapolate back to the past, we come to a point of infinite density. Can we extrapolate beyond this point? To give an idea of the range of values involved here, it is useful to define Planck's scales, which measure the length, time, and energy obtained by using three universal constants: h, Planck's constant; G, the gravitational constant; and c, the velocity of light. We then obtain Planck's length, $l = (Gh/c^3) \sim 10^{-33}$ cm, Planck's time on the order of 10^{-44}

seconds, and Planck's energy, corresponding to a high temperature on the order of 10^{32} degrees. It is plausible that these scales relate to the very early universe characterized by an extraordinarily brief time, a minuscule geometrical size, and an enormous energy. In this "Planck era," quantum effects are likely to play an essential role.[19] We have now arrived at the very limits of modern-day physics, where we are confronted with the fundamental problem of the quantization of gravity or, equivalently, of space-time. A general solution is still far from our grasp, but we may at least formulate a model that includes the role of Poincaré resonances and irreversibility at the very beginning of our universe. Let us now describe some of the steps that led us to this model.

We have noted that the Friedmann space-time interval can be written (when we consider the case of Euclidean three-dimensional geometry) as $ds^2 = \Omega^2(t)(dt_c^2 - dl^2)$, where t_c is the *conformal time*. This is the Minkowski space-time interval multiplied by the function Ω^2, which is called the *conformal factor*. Such conformal space-time intervals have remarkable features, including their conservation of the cone of light, for which $ds^2 = 0$. As Narlikar and others have stated, they are the natural starting point for quantum cosmology because they include the Friedmann universe as a special case.[20]

The conformal factor as a function of space-time relates to a field in the same way as do other fields such as the electromagnetic field. (Remember that a field is a dynamical system characterized by a well-defined energy and therefore a Hamiltonian). As shown by Robert Brout and his coworkers, this factor has a unique quality in that it corresponds to a negative energy (that is, its energy is unbounded from below), while the energy of any given matter field is positive. As a result, the gravitational field described by the conformal factor may play the role of a reservoir of negative energy from which the energy to create matter is extracted.[21]

This is the theoretical basis of the "free lunch" model, where the total energy (gravitational field plus matter) is conserved, while the gravitational energy is transformed into matter. Brout and his colleagues have proposed a mechanism for this extraction of positive energy. In addition to the conformal field, they have introduced a matter field, and demonstrated that Einstein's equations lead to a cooperative process involving the simultaneous appearance of matter and a curved space-time starting from the Minkowski space-time (containing zero gravitational and mass energy). Their model shows that such a cooperative process causes the exponential growth of the radius of the universe over the course of time. (This is known as the de Sitter universe.)

These conclusions are intriguing inasmuch as they indicate the possibility of an *irreversible* process transforming gravitation into matter. They also focus our attention on the preuniverse stage, the Minkowski vacuum, which is the starting point for irreversible transformations. It is important to note that this model does not describe creation *ex nihilo*. The quantum vacuum is already endorsed by the universal constants, and it is assumed that we can ascribe to them the values they have today.

The birth of our universe is no longer associated with a singularity, but rather with an instability that is analogous to a phase transition or bifurcation. However, this theory still presents a number of vexing problems. Brout et al. have used a semiclassical approximation in which the matter field is quantized while the conformal field is treated classically. This situation is highly unlikely in Planck's era, where quantum effects play an essential role.

Edgar Gunzig and Pasquale Nardone have asked why this process does not occur on a continuous basis if the quantum vacuum associated with a flat geometrical background is indeed unstable in the presence of gravitational interactions. They have demonstrated that in this semiclassical approximation, we need an initial fluctuation of a cloud of heavy particles of mass on the order of 50 Planck masses ($\sim 50.10^{-5}g$) in order to start the process.[22]

These results can be incorporated into a macroscopic thermodynamic approach, where the universe has to be treated as an open system. Thus, we can observe matter and energy being created at the expense of gravitational energy (see Figure 8.4). This compels us to make a number of modifications to the first law of thermodynamics, where there is now a source of matter-energy leading to a change in the definition of quantities such as pressure.* Since entropy is specifically associated with matter, the transformation of space-time into matter corresponds to a dissipative, irreversible process producing entropy. The inverse process, which would transform matter into space-time, is impossible. The birth of our universe would thus be the result of a burst of entropy.

The interaction of the gravitational and matter fields leads to divergences arising from brief times and short distances that correspond in quantum theory to high values of energy and momentum. These so-called "ultraviolet" divergences are the object of a number of interesting in-

---

*The "creation" pressure is negative. Therefore, an often-quoted theorem of Hawking and Penrose showing that the universe starts with a singularity and involves positive pressure is not applicable.

Figure 8.4

Matter Is Created at the Expense of the Gravitational Field

In this simple model, the universe would have no stable ground state.

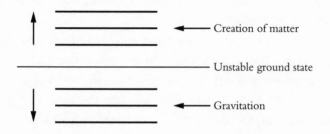

vestigations that have led to a procedure known as the renormalization program, which has proved to be quite successful. Still, certain difficulties remain. There is a striking analogy between field theory and the thermodynamic situation discussed in earlier chapters. Here again, we are dealing with persistent interactions that neither start nor stop, and we therefore have to go beyond Hilbert space.

Although this new field theory is still in the making, its main conclusion is reasonable: There may be no stable ground state at the cosmological level, since the conformal factor reaches lower energies as it creates matter. While this line of research continues to be pursued, the two concepts emphasized in this book, irreversibility and probability, clearly form an important part of this approach. Universes appear at sites where the amplitudes of the gravitational and matter fields have high values. The places and the times where this occurs have only a statistical meaning, as they are associated with quantum fluctuations of the fields. This description applies not only to our universe, but also to the meta-universe, the medium in which individual universes are born. In our view, here again we have an example of Poincaré resonances similar to that of the decay of an excited atom. In this case, however, the decay process creates not photons but universes! Even before *our* universe was created, there was an arrow of time, and this arrow will go on forever.

Of course, thus far we have only a simplified model. Einstein's dream of a unified theory that would include all interactions remains alive today.[23] Nonetheless, such a theory would have to take into account the time-oriented character of the universe as associated with its birth and subsequent evolution. This can be achieved only if certain fields (such as gravitation) play different roles from others (such as matter). In other words, unification is not enough. We need a more dialectical view of nature.

Questions concerning the origins of time will probably always be with us. But the idea that time has no beginning—that indeed time precedes the existence of our universe—is becoming more and more plausible.

Chapter 9

A NARROW PATH

I

It has often been suggested that irreversibility has a cosmological origin associated with the birth of our universe. It is true that cosmology is needed to explain why the arrow of time is universal, but irreversible processes did not cease with the creation of our universe; they still go on today, on all levels including geological and biological evolution. Although the dissipative structures introduced in Chapter 2 are routinely observed in the laboratory as well as in large-scale processes occurring in the biosphere, irreversibility can be fully understood only in terms of a microscopic description that was traditionally identified with classical and quantum mechanics. This requires a new formulation of the laws of nature that is no longer based on certitudes, but rather possibilities. In accepting that the future is not determined, we come to the end of certainty. Is this an admission of defeat for the human mind? On the contrary, we believe that the opposite is true.

183

The Italian author Italo Calvino has written a delightful collection of stories, *Cosmocomics,* in which individuals living in a very early stage of our universe gather together to remember the terrible time when the universe was so small that their bodies filled it completely.[1] What would have been the history of physics if Newton had been a member of this community? He would have observed the birth and decay of particles, the mutual annihilation of matter and antimatter. From the start, the universe would have appeared as a thermodynamic system far from equilibrium, with instabilities and bifurcations.

It is true that today we can isolate simple dynamical systems and verify the laws of classical and quantum mechanics. Still, they correspond to idealizations applicable to stable dynamical systems within a universe that is a giant thermodynamic system far from equilibrium, where we find fluctuations, instabilities, and evolutionary patterns at all levels. On the other hand, certainty has long been associated with a denial of time and creativity. It is interesting to consider this conundrum in its historical context.

II

How can we reach certainty? This is the question that lies at the heart of the work of René Descartes. In his thought-provoking book *Cosmopolis,* Stephen Toulmin attempts to clarify the circumstances that led Descartes on this quest.[2] He describes the tragic situation of the seventeenth century, a time of political instability and war between Catholics and Protestants in the name of religious dogma. It was in the midst of this strife that Descartes began his search for a different kind of certainty that all humans, independent of their religions, could share. This

led him to his famous *cogito,* the foundation of his philosophy, as well as his conviction that science based on mathematics was the only way to reach such certainty. Descartes'
views, which proved to be immensely successful, influenced Leibniz's concept of the laws of nature discussed in
Chapter 1. (Leibniz also wanted to create a language that
would heal the divisions among religions and bring about
the end of religious wars.) Descartes' pursuit of certainty
found its concrete realization in Newton's work, which has
remained the model for physics for three centuries.

Toulmin's analysis reveals a remarkable parallel between
the historical circumstances surrounding Descartes' quest
for certainty and those of Einstein's. For Einstein as well,
science was a means of avoiding the turmoil of everyday
existence. He compared scientific activity to the "longing
that irresistibly pulls the town-dweller away from his noisy,
cramped quarters and toward the silent high mountains."[3]

Einstein's view of the human condition was profoundly
pessimistic. He had lived through a particularly tragic period in human history spanning the rise of fascism and
anti-Semitism and two world wars. His vision of physics
has been defined as the ultimate triumph of human reason
over a violent world, separating objective knowledge from
the domain of the uncertain and the subjective.

But is science as conceived by Einstein—an escape from
the vagaries of human existence—still the science of today?
We cannot desert the polluted towns and cities for the high
mountains. We have to participate in the building of to-
morrow's society. In the words of Peter Scott, "The world,
our world, tries ceaselessly to extend the frontiers of the
knowable and the valuable, to transcend the givenness of
things, to imagine a new and better world."[4]

Science began with the Promethean affirmation of the

power of reason, but it seemed to end in alienation—a negation of everything that gives meaning to human life. Our belief is that our own age can be seen as one of a quest for a new type of unity in our vision of the world, and that science must play an important role in defining this new coherence.

As we mentioned in Chapter 8, at the end of his life, Einstein was offered a collection of essays that included a contribution by the great mathematician Kurt Gödel. In his answer to Gödel, he rejected his idea of a possible equivalence between past and future. For Einstein, no matter how great the temptation of the eternal, accepting the idea of traveling back in time was a denial of the real world. He could not endorse Gödel's radical interpretation of his very own views.[5]

As Carl Rubino has noted, Homer's *Iliad* revolves around the problem of time as Achilles embarks on a search for something permanent and immutable:

> The wisdom of the *Iliad,* a bitter lesson that Achilles, its hero, learns too late, is that such perfection can be gained only at the cost of one's humanity: he must lose his life in order to gain this new degree of glory. For human men and women, for us, immutability, freedom from change, total security, immunity from life's maddening ups and downs, will come only when we depart this life, by dying, or becoming gods: the gods, Horace tells us, are the only living beings who lead secure lives, free from anxiety and change.[6]

Homer's *Odyssey* appears as the dialectical counterpart to the *Iliad.* Odysseus is fortunate enough to be given the choice between immortality, by remaining forever the lover of Calypso, and a return to humanity and ultimately old age and death. In the end, he chooses time over eternity, human fate over the fate of the gods.

Since Homer, time has been the central theme of literature. We find a reaction quite similar to that of Einstein in an essay by the great writer Jorge Luis Borges entitled "A New Refutation of Time." After describing the doctrines that make time an illusion, he concludes: "And yet, and yet . . . denying temporal succession, denying the self, denying the astronomical universe, are apparent desperations and secret consolations. . . .Time is the substance I am made of. Time is a river which sweeps me along, but I am the river; it is the tiger which destroys me, but I am the tiger; it is a fire which consumes me; but I am the fire. The world, unfortunately, is real; I, unfortunately, am Borges."[7] Time and reality are irreducibly linked. Denying time may either be a consolation or a triumph of human reason. It is always a negation of reality.

The denial of time was a temptation for both Einstein the scientist and Borges the poet. Einstein repeatedly stated that he had learned more from Fyodor Dostoyevsky than from any physicist. In a letter to Max Born in 1924, he wrote that if he were forced to abandon strict causality, he "would rather be a cobbler, or even an employee in a gaming house, than a physicist."[8] In order to be of any value at all, physics had to satisfy his need to escape the tragedy of the human condition. "And yet, and yet," when confronted by Gödel with the extreme consequences of his quest, the denial of the very reality that physicists endeavor to describe, Einstein recoiled.

We can certainly understand Einstein's refusing chance as the only answer to our questions. What we have tried to follow is indeed a narrow path between two conceptions that both lead to alienation: a world ruled by deterministic laws, which leaves no place for novelty, and a world ruled by a dice-playing God, where everything is absurd, acausal, and incomprehensible.

We have attempted to make this book a journey along the narrow path, and thereby illustrate the role of human creativity in science. Strangely enough, this creativity is often undervalued. We all realize that if Shakespeare, Beethoven, or van Gogh had died soon after birth, no one else would ever have achieved what they did. Is this also true for scientists? Would someone else not have discovered the classical laws of motion if there had been no Newton? Did the formulation of the second law of thermodynamics depend entirely on Clausius? There is some truth in the contrast between artistic and scientific creativity. Science is a collective enterprise. In order to be acceptable, the solution to a scientific problem must satisfy exacting criteria and demands. These constraints, however, do not eliminate creativity. They provoke it.

The formulation of the time paradox is itself an extraordinary feat of human creativity and imagination. If science had been restricted to empirical facts, how could it ever have dreamed of denying the arrow of time? The elaboration of time-symmetrical laws was not achieved merely by introducing arbitrary simplifications. It combined empirical observations with the creation of theoretical structures. This is why the resolution of the time paradox could not be accomplished by a simple appeal to common sense or *ad hoc* modifications of the laws of dynamics. It was not even a matter of simply identifying the weaknesses in the classical edifice. In order to make fundamental progress, we needed to introduce new physical concepts, such as deterministic chaos and Poincaré resonances, and new mathematical tools to turn these weaknesses into strengths. In our dialogue with nature, we transform what first appear as obstacles into original conceptual structures providing fresh insights into the relationship between the knower and the known.

What is now emerging is an "intermediate" description that lies somewhere between the two alienating images of a deterministic world and an arbitrary world of pure chance. Physical laws lead to a new form of intelligibility as expressed by irreducible probabilistic representations. When associated with instability, whether on the microscopic or macroscopic level, the new laws of nature deal with the possibility of events, but do not reduce these events to deductible, predictable consequences. This delimitation of what can and cannot be predicted and controlled may well have satisfied Einstein's quest for intelligibility.

As we follow along the narrow path that avoids the dramatic alternatives of blind laws and arbitrary events, we discover that a large part of the concrete world around us has until now "slipped through the meshes of the scientific net," to use Alfred North Whitehead's expression.[9] We face new horizons at this privileged moment in the history of science, and it is our hope that we have been able to communicate this conviction to our readers.

NOTES

Acknowledgments

1. I. Prigogine and I. Stengers, *Entre le Temps et l'Eternité* (Paris: Librairie Arthème Fayard, 1988 (2nd ed., Paris, Flammarion, 1992).
2. I. Prigogine and I. Stengers, *Das Paradox der Zeit* (Munich: R. Piper & Co. Verlag, 1993); I. Prigogine and I. Stengers, *Time, Chaos and Quantum Theory* (Moscow: Ed. Progress, 1994).
3. I. Prigogine, *La Fin des Certitudes* (Paris: Odile Jacob, 1996).
4. I. Prigogine and I. Stengers, *Order Out of Chaos* (New York: Bantam Books, 1984); I. Prigogine, *From Being to Becoming* (San Francisco: W. H. Freeman, 1980).

Introduction

1. K. R. Popper, *The Open Universe: An Argument for Indeterminism* (Cambridge: Routledge, 1982), p. xix.
2. W. James, "The Dilemma of Determinism," in *The Will to Believe* (New York: Dover, 1956).
3. G. Gigerenzer, Z. Swijtink, T. Porter, J. Daston, J. Beatty, and L. Krüger, *The Empire of Chance* (Cambridge: Cambridge University Press, 1989), p. xiii.
4. See L. Krüger, J. Daston, and M. Heidelberger, eds., *The Probabilistic Revolution* (Cambridge, Mass.: MIT Press, 1990), 1:80.
5. Gigerenzer et al., *Empire of Chance*.
6. Popper, *Open Universe*.
7. R. Tarnas, *The Passion of the Western Mind* (New York: Harmony, 1991), p. 443.

8. I. Leclerc, *The Nature of Physical Existence* (London: Allen and Unwin; New York: Humanities Press, 1972).
9. J. Bronowski, *A Sense of the Future* (Cambridge, Mass.: MIT Press, 1978), p. ix.
10. S. Hawking, *A Brief History of Time: From the Big Bang to Black Holes* (New York: Bantam Books, 1988).

Chapter 1. Epicurus' Dilemma

1. For Epicurus, see J. Barnes, *The Presocratic Philosophers* (London: Routledge, 1989). He probably had in mind the Stoics, who believed in a kind of determinism.
2. For Lucretius, see Titus Lucretius Carus, *De Natura Rerum,* ed. C. Bailey (Oxford: Oxford University Press, 1947).
3. K. R. Popper, *The Open Society and Its Enemies* (Princeton, N.J.: Princeton University Press, 1963).
4. For Parmenides, see Barnes, *Presocratic Philosophers.*
5. Plato, *The Sophist* (New York: Garland, 1979).
6. J. Wahl, *Traité de Métaphysique* (Paris: Payot, 1968).
7. P. S. Laplace, *Oeuvres Complètes de Laplace* (Paris: Gauthier-Vilars, 1967).
8. G. von Leibniz, *Discourse on Metaphysics and Other Essays,* ed. D. Garber and R. Ariew (Indianapolis: Hackett, 1991).
9. J. Needham, *Science and Society in East and West: The Grand Titration* (London: Allen and View, 1969).
10. For the Einstein-Tagore correspondence, translated by A. Robinson, see K. Dutta and A. Robinson, *Rabindranath Tagore* (London: Bloomsbury, 1995).
11. Popper, *Open Universe,* loc. cit.
12. H. Bergson, *Oeuvres* (Paris: Presses Universitaires de France, 1959), p. 1331.
13. James, *Dilemma of Determinism,* loc. cit.
14. J. Searle, "Is There a Crisis in American Higher Education?" *Bulletin of the American Academy of Arts and Sciences* 46, no. 4 (January 1993): 24.
15. *Scientific American* 271, no. 4 (October 1994).
16. S. Weinberg, in ibid., p. 44.
17. Hawking, *Brief History of Time,* loc. cit.
18. R. Descartes, *Méditations métaphysiques* (Paris: J. Vrin, 1976).
19. R. Penrose, *The Emperor's New Mind* (Oxford: Oxford University Press, 1990), pp. 4–5.

20. A. N. Whitehead, *Process and Reality,* ed. D. Griffin and D. Sherborne, corrected ed. (New York: Macmillan, 1978).

21. C. P. Snow, *The Two Cultures and the Scientific Revolution. The Two Cultures and a Second Look.* (Cambridge: Cambridge University Press, 1964).

22. R. J. Clausius, *Ann. Phys.* 125 (1865): 353; Prigogine and Stengers, *Order Out of Chaos,* loc. cit.

23. A. S. Eddington, *The Nature of the Physical World* (Ann Arbor: University of Michigan Press, 1958).

24. See Prigogine, *From Being to Becoming.*

25. H. Poincaré, "La Mécanique et l'expérience," in *Revue de Métaphysique et Morale* 1 (1893): 534–537, and *Leçons de Thermodynamique,* ed. J. Blondin (Paris: Herman, 1923).

26. For Zermelo, see S. Brush, *Kinetic Theory* (New York: Pergamon Press, 1962), vol. 2.

27. R. Smoluchowski, "Vorträge über die kinetische Theorie der Materie und Elektrizität," 1914, quoted in H. Weyl, *Philosophy of Mathematics and Natural Science* (Princeton, N.J.: Princeton University Press, 1949).

28. M. Gell-Mann, *The Quark and the Jaguar* (London: Little, Brown, 1994), pp. 218–220.

29. M. Planck, *Treatise on Thermodynamics* (New York: Dover, 1945).

30. M. Born, *The Classical Mechanics of Atoms* (New York: Ungar, 1960); quoted in M. Tabor, *Chaos and Integrability in Nonlinear Dynamics* (New York: Wiley, 1969).

31. Prigogine, *From Being to Becoming,* p. 213.

32. See H. Price, *Time's Arrow and Archimedes' Point: New Directions for the Physics of Time* (Oxford: Oxford University Press, 1996).

33. J. L. Lagrange, *Théorie des fonctions analytiques* (Paris: Imprimerie de la République, 1796).

34. Gell-Mann, *Quark and the Jaguar.*

35. L. Rosenfeld, "Unphilosophical Considerations on Causality in Physics," in *Selected Papers of Léon Rosenfeld,* ed. R. S. Cohen and J. J. Stachel, *Boston Studies in the Philosophy of Science,* vol. 21 (Dordrecht: Reidel, 1979), pp. 666–690.

36. Borel, quoted in L. Krüger, J. Daston, and M. Heidelberger, *Probabilistic Revolution.*

37. J. W. Gibbs, *Elementary Principles in Statistical Mechanics* (New York: Scribner's, 1902).

38. H. Poincaré, *The Value of Science* (New York: Dover, 1958).

39. B. Mandelbrot, *The Fractal Geometry of Nature* (San Francisco: W.H. Freeman, 1983).
40. H. Poincaré, *New Methods of Celestial Mechanics,* ed. D. Goroff (American Institute of Physics, 1993).
41. M. Born, quoted in M. Tabor, *Chaos and Integrability in Nonlinear Dynamics,* p. 105.
42. Tabor, *Chaos and Integrability.*
43. M. Jammer, *The Philosophy of Quantum Mechanics* (New York: Wiley-Interscience, 1974); A. I. M. Rae, *Quantum Physics: Illusion or Reality?* (Cambridge: Cambridge University Press, 1986).
44. P. Davies, *The New Physics:A Synthesis* (Cambridge: Cambridge University Press, 1989), p. 6.
45. Quoted by K. V. Laurikainen, *Beyond the Atom:The Philosophical Thought of Wolfgang Pauli* (Berlin: Springer Verlag, 1988), p. 193.
46. Cl. George, I. Prigogine, and L. Rosenfeld, "The Macroscopic Level of Quantum Mechanics," *Kong. Danske Viden. Selskab Matematisk-fysiske Medd.* 38 (1972): 1–44.
47. See, e.g., W. G. Unruh and W. H. Zurek, "Reduction of a Wavepacket in Quantum Brownian Motion," *Phys. Rev.* 40 (1989): 1070.
48. J. S. Bell, *Speakable and Unspeakable in Quantum Mechanics* (Cambridge: Cambridge University Press, 1989).
49. Gell-Mann, *Quark and the Jaguar.*
50. G. C. Ghirardi, A. Rimini, and T. Weber, *Phys. Rev.* D34 (1986): 470.
51. B. d'Espagnat, *Conceptual Foundations of Quantum Theory,* Benjamin, California, 1976.
52. See I. Farquhar, *Ergodic Theory* (London: Interscience Publishers, 1964.)
53. J. von Neumann, *Mathematical Foundations of Quantum Mechanics* (Princeton, N.J.: Princeton University Press, 1955).
54. Cohen, *Probabilistic Revolution.*
55. H. Poincaré, *Science and Hypothesis* (New York: Science Press, 1921).

Chapter 2. Only an Illusion?

1. I. Prigogine, *Bull. Acad. Roy. Belgique* 31 (1945): 600. See also *Etude thermodynamique des phénomènes irrèversibles* (Liège: Desoer, 1947).

2. Lagrange, *Théorie des fonctions analytiques.*
3. Hawking, *Brief History of Time.*
4. Bergson, *L'Evolution créatrice,* in *Oeuvres,* p. 784.
5. Ibid., p. 1344.
6. Poincaré, *Science and Hypothesis.*
7. Whitehead, *Process and Reality.*
8. Eddington, *Nature of the Physical World.*
9. T. De Donder and P. Van Rysselberghe, *Affinity* (Menlo Park, Calif.: Stanford University Press, 1967); I. Prigogine, *Introduction to Thermodynamics of Irreversible Processes,* 3rd ed. (New York: Wiley, 1967).
10. G. N. Lewis, *Science* 71 (1930): 570.
11. E. Schrödinger, *What Is Life?* (Cambridge: Cambridge University Press, 1945).
12. I. Prigogine, *Bull. Acad. Roy. Belgique* 3, (1945): 600.
13. L. Onsager, *Phys. Rev.* 37 (1931): 405; 38 (1931): 2265. The proof of this theorem involves the celebrated Onsager reciprocity relations.
14. P. Glansdorff and I. Prigogine, *Thermodynamic Theory of Structure, Stability and Fluctuations* (New York: Wiley-Interscience, 1971).
15. G. Nicolis and I. Prigogine, *Exploring Complexity* (San Francisco: Freeman, 1989).
16. Ibid.
17. For a review of oscillatory reactions, see *Chemical Waves and Patterns,* ed. R. Kapral and K. Showalter (Newton, Mass: Kluwer, 1995).
18. For a review of nonequilibrium spatial structures, see Special Issue of *Physica A* 213, nos. 1–2, "Inhomogeneous Phases and Pattern Formation," ed. J. Chanau and R. Lefever (North-Holland, 1995).
19. A. M. Turing, *Phil. Trans. Roy. Soc. London,* Ser. B, 237 (1952): 37.
20. Nicolis and Prigogine, *Self-Organization* and *Exploring Complexity.*
21. Nicolis and Prigogine, *Exploring Complexity;* Prigogine, *From Being to Becoming.*
22. C. K. Biebracher, G. Nicolis, and P. Schuster, *Self-Organization in the Physico-Chemical and Life Sciences,* Report EUR 16546 (European Commission, 1995).

Chapter 3. From Probability to Irreversibility

1. Prigogine, *From Being to Becoming.*
2. P. and T. Ehrenfest, *Conceptual Foundations of Statistical Mechanics* (Ithaca, N.Y.: Cornell University Press, 1959).
3. A. Bellemanns and J. Orban, *Phys. Letters* 24A (1967): 620.
4. I. Prigogine, *Nonequilibrium Statistical Mechanics* (New York: Wiley, 1962); R. Balescu, *Equilibrium and Non Equilibrium Statistical Mechanics* (New York: Wiley, 1975); P. Resibois and M. De Leener, *Classical Kinetics of Fluids* (New York: Wiley, 1977).
5. A. Lasota and M. Mackey, *Probabilistic Properties of Deterministic Systems* (Cambridge: Cambridge University Press, 1985).
6. Jan von Plato, *Creating Modern Probability: Its Mathematics, Physics, and Philosophy in Historical Perspective* (Cambridge, Mass: Cambridge University Press, 1994).
7. D. Ruelle, *Phys. Rev. Letters* 56 (1986): 405; *Commun. Math Phys.* 125 (1989): 239; H. Hasegawa and W. C. Saphir, *Phys. Rev. A* 46 (1992): 7401; H. Hasegawa and D. Driebe, *Phys. Rev. E* 50 (1994): 1781; P. Gaspard, *J. of Physics A* 25 (1992): L483; I. Antoniou and S. Tasaki, *J. of Physics A: Math. Gen.* 26 (1993): 73; *Physica A* 190 (1992): 303.
8. I. Prigogine, *Les Lois du Chaos* (Paris: Flammarion, 1994), and *Le leggi del caos* (Rome: Laterza, 1993).

Chapter 4: The Laws of Chaos

1. Hasegawa and Saphir, *Phys. Rev. A* 46 (1992): 7401; Hasegawa and Driebe, *Phys. Rev. E* 50 (1994): 1781; P. Collet and J. Eckman, *Iterated Maps on the Interval as Dynamical Systems* (Boston: Birckhauser, 1980); P. Shields, *The Theory of Bernoulli Shifts* (Chicago: University of Chicago Press, 1973).
2. P. Duhem, *La théorie physique. Son objet. Sa structure* (reprint, Paris: Vrin, 1981), vol. 2.
3. Hasegawa and Saphir, *Phys. Rev. A* 46 (1992): 7401; Hasegawa and Driebe, *Phys. Rev. E* 50 (1994): 1781; Gaspard, *Journal of Physics* 25 (1992): L483; Antoniou and Tasaki, *Journal of Physics A: Math. Gen.* 26 (1993): 73.
4. Ibid.
5. Mandelbrot, *The Fractal Geometry of Nature;* P. and T. Ehrenfest, *Conceptual Foundations of Statistical Mechanics.*

6. Nicolis and Prigogine, *Exploring Complexity;* Prigogine, *From Being to Becoming.*
7. See, e.g., F. Riesz and B. Sz-Nagy, *Functional Analysis* (New York: Dover, 1991).
8. Prigogine, *From Being to Becoming;* V. Arnold and A. Avez, *Ergodic Problems of Classical Mechanics* (New York: Benjamin, 1968).
9. Hasegawa and Saphir, *Phys. Rev. A* 46 (1992): 7401; Hasegawa and Driebe, *Phys. Rev. E* 50 (1994): 1781; Gaspard, *Journal of Physics* 25 (1992): L483; Antoniou and Tasaki, *Journal of Physics A: Math. Gen.* 26 (1993): 73.
10. P. Gaspard, *Physics Letters A* 168 (1992): 13, and *Chaos* 3 (1993): 427; H. Hasegawa and D. Driebe, *Physics Letters A* 168 (1992): 18, and *Phys. Rev. E* 50 (1994): 1781; H. Hasegawa and E. Luschei, "Exact Power Spectrum for a System of Intermittent Chaos," *Physics Letters A* 186 (1994): 193.

Chapter 5: Beyond Newton's Laws

1. T. Petrosky and I. Prigogine, "Alternative Formulation of Classical and Quantum Dynamics for Non-Integrable Systems," *Physica A* 175 (1991); T. Petrosky and I. Prigogine, "Poincaré Resonances and the Limits of Trajectory Dynamics," *PNAS* 90 (1993): 9393; T. Petrosky and I. Prigogine, "Poincaré Resonances and the Extension of Classical Dynamics," *Chaos, Solitons and Fractals* 5 (1995).
2. See any text on Fourier series.
3. Prigogine, *Nonequilibrium Statistical Mechanics.*
4. See Petrosky and Prigogine, "Poincaré Resonances."
5. See S. G. Brush, *Kinetic Theory* (Oxford: Pergamon Press, 1972), vol. 3.
6. See Y. Pomeau and P. Résibois, *Physics Reports* 19, 2 (1975): 63.
7. T. Petrosky and I. Prigogine, "New Methods in Dynamics and Statistical Physics" (forthcoming).
8. Prigogine, *Nonequilibrium Statistical Mechanics*; see also the citations in note 1 of this chapter.

Chapter 6: A Unified Formulation of Quantum Theory

1. R. Penrose, *Shadows of the Mind* (Oxford: Oxford University Press, 1994), chap. 5.
2. P. Davies, *The New Physics;* Rae, *Quantum Physics.*
3. J. C. von Neumann, *Mathematical Foundations of Quantum Theory.*
4. T. Petrosky and I. Prigogine, "Quantum Chaos, Complex Spectral Representations and Time-Symmetry Breaking," *Chaos, Solitons and Fractals* 4 (1994): 311; T. Petrosky and I. Prigogine, *Physics Letters A* 182 (1993): 5; T. Petrosky, I. Prigogine, and Z. Zhang (forthcoming).
5. K. R. Popper, *Quantum Theory and the Schism in Physics* (Totowa, N.J.: Rowman and Littlefield, 1982).
6. The standard text is by P. A. M. Dirac, *The Principles of Quantum Mechanics* (Oxford: Oxford University Press, 1958).
7. M. Jammer, *The Philosophy of Quantum Mechanics* (New York: John Wiley, 1974).
8. A. Eddington, *The Nature of the Physical World* (Ann Arbor: University of Michigan Press, 1958).
9. A. Böhm, *Quantum Mechanics* (Berlin: Springer, 1986); A. Böhm and M. Gadella, *Dirac Sets, Gamov Vectors and Gelfand Triplets* (Berlin: Springer, 1989); G. Sudarshan, *Symmetry Principles at High Energies,* ed. A. Perlmutter et al. (San Francisco: Freeman, 1966); G. Sudarshan, C. B. Chiu, and V. Gorini, *Physical Review D* 18 (1978): 2914.
10. Petrosky and Prigogine, "Quantum Chaos;" T. Petrosky and Z. Zhang (forthcoming).
11. Petrosky and Prigogine, "Quantum Chaos" and *Physics Letters;* Petrosky, Prigogine, and Zhang (forthcoming).
12. N. Bohr, "The Solvay Meeting and the Development of Quantum Physics," in *La Théorie quantique des champs* (New York: Interscience, 1962).

Chapter 7: Our Dialogue with Nature

1. V. Nabokov, *Look at the Harlequins* (New York: McGraw-Hill, 1974).
2. Prigogine and Stengers, *Entre le Temps et l'Eternité.*
3. N. Bohr, *Atomic Physics and Human Knowledge* (New York: Wiley, 1958).

4. A. I. M. Rae, *Quantum Physics.*

5. W. Heisenberg, *The Physical Principles of the Quantum Theory* (Chicago: University of Chicago Press, 1930).

6. See Nicolis and Prigogine, *Exploring Complexity.*

7. S. J. Gould, *Scientific American* 271, no. 4 (October 1994): 84.

8. Ibid.

Chapter 8: Does Time Precede Existence?

1. J. Wheeler, quoted in H. Pagels, *Perfect Symmetry* (New York: Bantam Books, 1986), p. 165.

2. L. D. Landau and E. M. Lifschitz, *The Classical Theory of Fields* (London: Pergamon Press, 1959).

3. P. Davies, *About Time* (London: Viking, 1995).

4. H. Minkowski, *The Principle of Relativity: Original Papers* (Calcutta: University of Calcutta, 1920).

5. A. Einstein, *Correspondence Einstein-Michele Besso 1903–1955* (Paris: Hermann, 1972).

6. *Albert Einstein: Philosopher-Scientist,* ed. P. A. Schlipp (Evanston, Ill.: Library of Living Philosophers, 1949).

7. H. Bondi, *Cosmology* (Cambridge: Cambridge University Press, 1960).

8. See J. V. Narlikar and T. Padmanabhan, *Gravity, Gauge Theory and Quantum Cosmology* (Dordrecht: Reidel, 1986).

9. I. Antoniou and B. Misra, *Journal of Theoretical Physics* 31 (1992): 119.

10 V. Fock, *The Theory of Space, Time and Gravitation* (New York: Pergamon Press, 1959).

11. Hawking, *Brief History of Time.*

12. P. A. M. Dirac, *Rev. Mod. Phys.,* 21 (1949): 392; D. J. Currie, T. F. Jordan, and E. C. G. Sudarshan, *Rev. Mod. Phys.,* 35 (1962): 350; R. Balescu and T. Kotera, *Physica* 33 (1967): 558; U. Ben Ya'acov, *Physica.*

13. Bondi, *Cosmology.*

14. See the excellent account by S. Weinberg, *The First Three Minutes: A Modern View of the Origin of the Universe* (New York: Basic Books, 1977).

15. See Alpher and Herman, in *Nature* 162 (1948): 774, and *Physical Review* 75, no. 7 (1949): 1089.

16. See E. P. Tryon, in *Nature* 266 (1973): 396.

17. See, for a general account, S. Weinberg, *Gravitation and Cosmology: Principles and Applications of the General Theory of Relativity* (New York; Wiley, 1972).
18. Ibid.
19. See J. V. Narlikar and T. Padmanabham, *Gravity.*
20. Narlikar and Padmanabhan, *Gravity.*
21. R. Brout, F. Englert, and E. Gunzig, *Ann. Phys.* 115 (1978): 78; *General Relativity and Gravitation* 10 (1979): 1; R. Brout et al., *Nuclear Physics B* 170 (1980): 228; E. Gunzig and P. Nardone, *Physics Letters B* 188 (1981): 412, and also in *Fundamentals of Cosmic Physics* 11 (1987): 311.
22. E. Gunzig, J. Géheniau, and I. Prigogine, *Nature* 330 (1987): 621; I. Prigogne, J. Géheniau, E. Gunzig, and P. Nardone, *Proc. Nat. Acad. Sci. USA* 85 (1988): 1428.
23. S. Weinberg, *Dreams of a Final Theory* (New York: Pantheon Books, 1992).

Chapter 9: A Narrow Path

1. I. Calvino, *Cosmicomics,* trans. W. Weaver (New York: Harcourt, Brace & World, 1969).
2. S. Toulmin, *Cosmopolis* (Chicago: Chicago University Press, 1990).
3. A. Einstein, *Ideas and Opinions* (New York: Crown, 1954), p. 225.
4. P. Scott, *Knowledge, Culture and the Modern University,* 75th Jubilee of the Rijksuniversiteit (Groningen, Holland, 1984).
5. *Albert Einstein: Philosopher-Scientist.*
6. Carlo Rubino, unpublished.
7. J. L. Borges, "A New Refutation of Time," *Labyrinths,* Penguin Modern Classics (Harmondsworth: Penguin Books, 1970), p. 269.
8. A. Einstein and M. Born, *The Born-Einstein Letters* (New York: Walker, 1971), p. 82.
9. A. N. Whitehead, *Process and Reality.*

GLOSSARY

anthropic principle The idea that the conditions of the universe are explained by the fact that we are here to observe them.

bifurcation The branching of a solution into multiple solutions as a system parameter is varied.

big bang The initial event of our universe, described as an explosive creation of matter and energy from a point.

chaos The behavior of systems in which close trajectories separate exponentially in time.

clinamen The idea, due to Epicurus, that an element of chance is needed to account for the deviation of material motion from rigid predetermined evolution.

coarse graining The averaging of dynamics over finite regions of phase space.

collapse of the wave function The extradynamical element needed in orthodox quantum theory for the wave function, representing potentialities, to yield an actual state.

degrees of freedom The number of independent variables needed to specify the configurational state of a system. A single particle in three-dimensional space has three degrees of freedom.

determinism The viewpoint that evolution is governed by a set of rules that, from any particular initial state, can generate one and only one sequence of future states.

deterministic chaos Chaotic behavior arising from an entirely deterministic evolution law.

Dirac delta function The mathematical object, introduced by Dirac, which may be considered a function defined as infinity at one point and zero everywhere else.

dissipative structure Spatiotemporal structures that appear in far-from-equilibrium conditions, such as oscillating chemical reactions or regular spatial structures.

eigenstate A state that when acted on by a given operator yields the same state multiplied by a number.

eigenvalue The number that an eigenstate is multiplied by after it is acted upon by the corresponding operator.

ensemble An imagined collection of identical systems with different initial conditions.

entropy A function of the state of the system that increases monotonically for isolated systems and reaches a maximum at thermodynamic equilibrium.

fractal The term coined by Benoît Mandelbrot for mathematical objects of noninteger dimension. For example, the length of the irregular coastline of a country increases as the scale used to measure it decreases, and so the coastline has a dimension between one and two.

Friedmann universe A cosmological model of an expanding universe based on the assumption of homogeneity and isotropy of the universe on large scales.

Gelfand space The function space containing both the generalized functions and the well-behaved functions they act on.

generalized function The class of mathematical objects to which the Dirac delta function belongs. A generalized function is not a regular mathematical function but is defined by how it acts on regular functions.

H-theorem Boltzmann's finding that a function (the H-function) involving the one-particle distribution function appears unidirectional in time behavior under evolution of a dilute gas of interacting particles.

Hamiltonian The energy of a dynamical system expressed in terms of its coordinates and momenta.

Heisenberg uncertainty principle The product of the accuracies by which the position and momentum of a quantum particle may be determined as limited by Planck's constant. Complete accuracy of either the position or the momentum implies complete indeterminacy of the other one.

Hilbert space The space of functions for which the integral of the square of the functions is well defined and finite. This is the function space that was used as the setting for orthodox quantum mechanics. It has been subsequently applied to classical mechanics and statistical mechanics.

KAM theory Describes the dynamical behavior of classes of nonintegrable systems. As the energy of a system is increased, chaotic behavior becomes more prevalent.

kinetic theory The study of the thermodynamic and transport properties of fluid and gas systems in terms of interparticle interactions.

large Poincaré system (LPS) A nonintegrable system due to Poincaré resonances taken in the thermodynamic limit so that its energy spectrum is continuous.

Laplace demon The entity imagined by Laplace that would be able, given the exact initial conditions, to calculate the precise future evolution of our universe.

Loschmidt's reversal paradox The argument, raised against the conclusions of Boltzmann, that since the equations of motion in an interacting particle system are reversible, one can consider reversing all the velocities in a system so that any time-oriented functions of the state of the system would then behave in an opposite manner.

Lyapunov exponent The rate of exponential separation of nearby trajectories in a chaotic system.

map A discrete-time dynamical process.

Markov process A process wherein the future evolution of a state depends only on the present state. For a continuous time system this means that the process is local in time, that is, there are no memory effects.

Newtonian dynamics The rules of evolution that form the core of classical physics and that, in pre-quantum era determinism, were believed to underlie all physical reality.

nonintegrable system An interacting system that cannot be transformed to noninteracting parts. If such a transformation can be performed, the system is integrable and the equations of motion can be trivially solved.

Perron–Frobenius operator The time evolution operator for probability distributions in discrete-time systems (maps).

phase space The abstract space of points in which the coordinates are the positions and velocities of the particles in an evolving system.

Planck era The universe just after the big bang characterized by the Planck scales, involving three fundamental constants of nature, h, c, and G.

Poincaré recurrence theorem The finding that the state of a closed system, as defined by the values of the positions and velocities of all the particles, will recur arbitrarily closely under time evolution of the system.

Poincaré resonances Coupling of degrees of freedom that lead to divergent expressions due to small denominators if there is resonance between them. The resonances may prohibit the solution of the equations of motion.

probability distribution function The function representing the relative weights of the systems or initial conditions distributed in an ensemble.

resonance The constructive interference that appears when two frequencies in a system are rationally related.

Ritz–Rydberg principle The frequency of spectral lines representing the difference between two energy levels.

second law of thermodynamics The principle that the entropy of an isolated system may only increase or remain constant under time evolution.

self-organization The choice between solutions appearing at a bifurcation point, determined by probabilistic laws. Far-from-equilibrium self-organization leads to increased complexity.

spectral decomposition The expression of an operator in terms of its eigenstates and eigenvalues in a given function space.

steady-state universe A cosmological model wherein the expansion of the universe is compensated by a continuous creation of matter.

thermodynamic limit The procedure of considering the number N of particles and the size V of a system becoming arbitrarily large while the concentration, $c = N/V$, remains finite and constant.

thermodynamics The study of the macroscopic properties of a system and their relations without regard to the underlying dynamics.

Turing structures Patterns in chemical systems arising from an interplay of reaction and diffusion processes; these are an example of dissipative structure.

INDEX

aging, 78, 125
Alper, Ralph A., 174
anthropic principle, 15–16
Antoniou, Ioannis, 169
approximation: dissipative structures not explained by, 73; evolution as due to, 23, 24–25;
  fundamental problems solved in
  terms of, 52; irreversibility as
  due to, 23, 24, 81, 91, 105
architecture, 60
Arnold, Vladimir Igorevich, 41
arrow of time: all having same orientation, 102, 162; Bernoulli
  maps introducing, 90, 96; constructive role of, 3; for dealing
  with intelligent life, 15; denial
  of, 1–2; dissipative structures re-
  quiring, 73; entropy as, 19; as
  eternal, 182; in evolving universe, 4; as fact imposed by observation, 74; hostility to
  concept of, 61–62; in macroscopic processes, 18; in nonequilibrium physics, 3;
  nonintegrable systems required
  for, 39; physics' denial of, 2; in
  realistic interpretation of quantum mechanics, 54; relegated to

phenomenology, 2, 3; as source
  of order, 26; space-time as consistent with, 172; in structure
  formation, 71; subjective interpretation of, 49
atomism, 9–10, 127
atoms, Bohr's theory of, 132–33,
  145

bacteria, 161
baker transformations (maps),
  96–105, *97;* approach to equilibrium in terms of, 102; and
  Bernoulli map, 90, 91; Bernoulli
  map compared with, 97–98,
  103–4; Bernoulli shift for representing, 99; as chaotic and deterministic, 101; eigenfunctions
  and eigenvalues for, 103; equivalence of individual and statistical
  description broken with, 103; as
  invertible, 101, 102; numerical
  simulation of, *98;* Perron-
  Frobenius operator with, 103;
  recurrence in, 99–101; spectral
  representation in, 103; successive
  iterations of, 98–99, *100;* time
  paradox associated with, 103; as
  time reversible, 101, 103

207